Queues Applied to Telecoms

Series Editor
Guy Pujolle

Queues Applied to Telecoms

Courses and Exercises

Toky Basilide Ravaliminoarimalalason
Falimanana Randimbindrainibe

iSTE

WILEY

First published 2022 in Great Britain and the United States by ISTE Ltd and John Wiley & Sons, Inc.

ISTE Ltd
27-37 St George's Road
London SW19 4EU
UK

www.iste.co.uk

John Wiley & Sons, Inc.
111 River Street
Hoboken, NJ 07030
USA

www.wiley.com

Library of Congress Control Number: 2022946967

British Library Cataloguing-in-Publication Data
A CIP record for this book is available from the British Library
ISBN 978-1-78630-904-4

Contents

Notations

Numerals

0 Vector with coordinates of zero
Matrix with coordinates of zero

1 Matrix whose coordinates are all equal to 1

$1_{(s,t)}$ Indicator function of the interval (s, t)

Latin lower-case letters

$a_{i,j,k}$ Any constant indexed by i, j and k

a_s Distribution of the number of customers who arrive while another customer is being served

b Offered traffic per free source

b_i Normalization constant for a queue i

c Length of any path
Average frequency of accepted requests

c_A Average frequency of all requests

c_R Average frequency of refused requests

d Any duration
Delay time

$\bar{d}$	Average duration of a call
det	Determinant of a matrix
d_i	Duration of the i-th call
$diag$	Diagonal function for creating a diagonal matrix
d_q	Waiting time for delayed requests
dt	Very short interval of time
$\mathbf{e}_i$	Vector whose i-th coordinate is equal to 1 and the others are zero
exp	Exponential with base *e*
$f(x)$	Probability density function for any random variable
f_X	Probability density function for a random variable *X*
$g(x)$	Probability density function for any random variable
h	Average holding time
$h(X)$	Entropy of a random variable *X*
h_i	Average duration of the i-th holding time
i	Any natural integer Any state of a Markov chain
$i.i.d$	Independent and identically distributed
i_k	Any state of a Markov chain
j	Any natural integer Any state of a Markov chain
k	Any natural integer
k_i	Any natural integer indexed by *i*
l	Any natural integer
lim	Limit

ln	Natural logarithm
mErl	Milli-Erlang, unit of measurement for traffic
min	Minimum
n	Any natural integer Number of service nodes in a delay system Number of nodes in a queueing network Number of resources
$n!$	Factorial of a natural integer n
n_i	Any natural integer indexed by i
o	Little-o function
p	Any real number between 0 and 1 Parameter of a geometric distribution
$p(i,j)$	Coordinate ij of the transition matrix of a Markov chain
p_i	Any real number between 0 and 1 Probability of leaving queue i
p_{ij}	Probability of being routed from queue i to queue j
q	Transition rate in a queueing network Dimensions of a queue
q_i	Parameter of the holding time distribution in state i Transition intensity from state i to state i
$\mathbf{q}_i$	Eigenvector of a matrix
q_{ij}	Transition intensity from state i to state j
r	Any time
s	Any time Number of simultaneously busy sources
t	Any time
t'	Any time

t_i	Any time indexed by i
t_i^+	Instant just after time t_i
w	Average delay time of requests that must effectively wait
$\mathbf{x}$	Vector for the number of customers Any column vector State of a queueing network
x_i	Number of customers in queue i
z	Variable of a generator function

Greek lower-case letters

ε	Infinitely small real positive number
θ	Period of time
λ	Parameter of an exponential distribution Intensity of an arrival process
λ_e	Intensity of an effective arrival process
λ_i	Parameter indexed by i for a random variable following an exponential distribution Intensity indexed by i of an arrival process Effective rate of arrival to queue i i-th eigenvalue of a matrix
μ	Parameter for an exponential duration distribution Inverse of the average duration of service
μ_i	Parameter indexed by i for an exponential duration distribution Inverse of the average duration of service for a queue i
μ_i^j	Inverse of the average duration of class-j service for a queue i
ν_i	Intensity of the arrival process to queue i
$\boldsymbol{\pi}$	Any row vector Stationary distribution

π_i i-th coordinate of the stationary distribution

$\dot{\boldsymbol{\pi}}(t)$ Derivative with respect to time of the function $\pi(t)$

$\boldsymbol{\pi}(\mathbf{x})$ Stationary distribution of state **x**

ρ Offered traffic
Load of a delay system

ρ_i Load of the i-th queue in a network

σ^2 Variance of a random variable

τ Inter-arrival of a punctual process

τ_n n-th inter-arrival of a punctual process

φ_i Capacity of queue i

Latin capital letters

A Distribution of inter-arrivals to a delay system in the Kendall notation
Offered traffic

$\tilde{A}$ Fictitious offered traffic

$\mathbf{A}$ Infinitesimal stochastic generator

A_p Lost traffic, refused traffic

AS Almost surely

B Distribution of durations of service of a delay system in the Kendall notation
Blocking probability

$B_{1,n}(A)$ Erlang loss formula, first Erlang formula, Erlang-B

$B_{2,n}(A)$ Erlang waiting formula, second Erlang formula, Erlang-C

B_K Normalization constant

B_k Probability of having k busy resources

$B_{n,N}$	Blocking probability for n resources and N sources
$\mathcal{B}(n,p)$	Binomial distribution with parameters n and p
C_S	Variation coefficient of the variable S
D	Deterministic distribution Period or duration of observation Random variable of durations of service Probability of waiting
D	Diagonal matrix of eigenvalues for eigendecomposition of a matrix
D_t	Duration of service already received by a customer at instant t
E	Loss probability
E_d	Loss probability due to source abandonment
$\mathcal{E}(n,\lambda)$	Erlang distribution with parameters n and λ
$E(X)$	Expected value of a random variable X
E_k	Erlang distribution with parameter k
$E_{n,N}$	Loss probability for n resources and N sources
Erl	Erlang, unit of measurement for traffic
$F_{n,N}$	Improvement factor for n resources and N sources
F_X^C	Inverse distribution function of the random variable X
G	General distribution
G'	Derivative of G
$G(z)$	Generator function of a stationary distribution
$G_X(z)$	Generator function of a random variable X
GI	General and independent distribution
H_k	Hyper-exponential distribution of order k
I	Number of queues

$\mathbf{I}$ Identity matrix

$\mathcal{L}_S$ Laplace transform

K Any natural integer
Total population of customers able to access a delay system
Total number of customers in a queueing network

K_π Normalization constant

M Markovian distribution

N Any random variable
Any natural integer
Capacity of a delay system
Number of sources

$\boldsymbol{N}$ Vector for the number of customers in a queueing network

$\mathbb{N}$ Set of natural integers

$\mathbb{N}^*$ Set of non-zero natural integers

$N(s,t)$ Counting measure of a process between instants s and t

$N_i(s,t)$ Counting measure of the i-th process between instants s and t

$N(t)$ Counting measure of a process up to instant t
Number of customers arriving in a queue during time t

N_a Number of calls during an observation period

N_i Number of customers in the i-th node of the queueing network

N_k Number of customers present in the system just after the departure of the k-th customer

N_q Number of customers in a system queue
Number of waiting requests

N_{qa} Number of waiting requests given there is a wait

N_s Number of customers in the system

NS The event "a request arrives"

N_t	Number of customers in the system at instant t
$\mathbf{P}$	Transition matrix for a Markov chain
$\tilde{\mathbf{P}}$	Transition matrix for a reversible Markov chain
$\mathbb{P}$	Probability function
$\mathcal{P}(\lambda)$	Poisson distribution with parameter λ
P_{ij}	Coordinate ij of the transition matrix for a Markov chain
$\tilde{P}_{ij}$	Coordinate ij of the transition matrix for a reversible Markov chain
$P_{ij}^{(t)}$	Coordinate ij of the matrix $\mathbf{P}^t$, where $\mathbf{P}$ has the coordinate P_{ij}
$\mathbf{Q}$	Transition matrix for the eigendecomposition of a matrix
R	Random variable for the number of simultaneously busy resources
$\mathbb{R}$	Set of real numbers
$\mathbb{R}^+$	Set of positive real numbers
S	Any sum of random variables Random variable for the number of simultaneously busy resources
S_n	Sum of n random variables
T	Half-life period Any time Duration of service Period of observation
T_i	Time spent by the system in state i before changing to another state
T_{ij}	Time spent by the system in state i before changing to state j
$\mathcal{T}$	Set of possible values that the time can take
T_i	Instants of appearance in a punctual process
T_q	Time spent by a customer in the system queue
T_s	Time spent by a customer in the system

T_t Punctual process indexed by time t

W Average delay time for all requests

X Real random variable
Number of requests

X_i Real random variable indexed by i

$X(t)$ Real random variable indexed by time t
Punctual process
Markov chain

$\tilde{X}(t)$ Reversible Markov chain

$\mathcal{X}$ Set of possible values a random variable can take
State space

$\{X_n\}$ Sequence or family of random variables, stochastic process in discrete time

$\{X(t)\}$ Family of random variables, stochastic process in continuous time

Y Carried traffic

Y_i Any event indexed by i
Random variable equal to the number of customers who arrived during the service of the i-th customer
Traffic carried individually by the i-th resource

Y_n Nominal traffic or maximal theoretical traffic
Carried traffic for n resources

Z Service discipline in a waiting space in the Kendall notation
Traffic peakedness

$\mathbb{Z}$ Set of relative integers

Greek capital letters

Δt Any interval of time

$\mathbf{\Pi}$ Matrix of the row coordinates $\boldsymbol{\pi}$

Φ Balance function

Special notations

■	QED, "which was to be demonstrated"
$:=$	By definition, equal to
$i \rightsquigarrow j$	State j can be reached from state i of a Markov chain
$i \leftrightsquigarrow j$	State j can be reached from state i and vice versa
$\binom{n}{k}$	Number of k combinations from a set of n elements
$\lfloor X \rfloor$	Floor function of X
$\lceil X \rceil$	Ceiling function of X

Preface

Queues are omnipresent in communication networks functioning in packet mode. We find them in every computer, router and radio access point. They are veritable funnels whose purpose is to maximize the use of network resources. It is at this level that policies for sharing among users through scheduling and the selective rejection of packets are established. When numerous data transfers share the same link, the system made up of the ensemble of files being transferred can itself be seen as a virtual queue, distributed from servers where the files being transferred are stored.

By extension, network models with circuit switches are considered a particular type of queue, given that there is no wait, since flows are simply allowed or rejected. In certain cases, as with mobile phone networks or call centers, an operator can set up a system that places calls on hold for more or less time, with the hope that a resource might soon become available. Formally speaking, we should thus speak of waiting and rejection lines; usage, however, dictates that the simpler term "queue" be used.

The theorization of teletraffic consists of studying the uses of resources made available to users in any system, and more specifically systems with delay times. For the case of telecommunications, it provides a means for planning a network, or the components of this network.

We can cite a number of objectives in the study of teletraffic: controlling planning that has already been developed, discovering resources that have become available, detecting potential problems, detecting network configuration problems or the programming of a central modifying network routing algorithms in a dynamic way, providing the indications that will serve for planning.

This book deals with the theoretical aspect of these queues: from the Poisson process, Markov chains and queueing systems to queuing networks. The study of the use of their resources is called the theory of teletraffic. This work also sets out the fundamentals of the theory of teletraffic by presenting the teletraffic of loss systems and that of queueing systems. Certain applications or explanations are more oriented towards the field of telecommunications. This book contains lessons and more than 60 exercises with solutions.

As a prerequisite, the reader should understand the basics of probability, which would be very useful to better understand the content of this book. Probability notations also facilitate an understanding of certain points listed in this book.

June 2022

PART 1

Typical Processes in Queues

1

The Poisson Process

Living without memory is an unforgettable experience.

Yolande Villemaire (1949–)

It is with this citation by the Quebecois novelist and poet Yollande Villemaire that we begin this chapter on the memoryless characteristics of the exponential function. We review this distribution, which is a probability tool frequently used in the study of queues. Using properties analogous to the exponential function, we can then introduce the Poisson[1] process in this chapter, describing the moments when random events occur, such as the arrival of customers in a queue.

1.1. Review of the exponential distribution

1.1.1. *Definitions*

DEFINITION 1.1.– *A real positive random variable X follows an exponential distribution with parameter $\lambda > 0$ if:*

$$\mathbb{P}(X > t) = e^{-\lambda t}, \ \forall t \in \mathbb{R}^+ \qquad [1.1]$$

The notation $\mathbb{P}$ indicates the probability function. In equation [1.1], the expression $\mathbb{P}(X > t)$ therefore designates the probability that the random variable X has a value that is superior to a positive real number t.

Just like any real random variable, it can be characterized by a probability density function.

1 Siméon Denis Poisson (1781–1840), a French mathematician, geometrician and physician.

Its density function is given by:

$$f_X(t) = \lambda e^{-\lambda t}, \ \forall t \in \mathbb{R}^+ \quad [1.2]$$

From the density function, we can also express the expected value [1.3] and the variance [1.4] of the random variable X following an exponential distribution with parameter λ:

$$E(X) = \int_0^{+\infty} t f_X(t)\, dt = \frac{1}{\lambda} \quad [1.3]$$

$$\sigma^2(X) = \int_0^{+\infty} \left(t - \frac{1}{\lambda}\right)^2 f_X(t)\, dt = \frac{1}{\lambda^2} \quad [1.4]$$

The exponential distribution is used in many applications in physics, biology, computer programing and, of course, telecommunications.

EXAMPLE.– In radioactivity, the lifetime of a radioactive particle can be represented by a random variable following an exponential distribution. The parameter for this distribution is the speed λ at which the particle ages (see Exercise 1.1).

It is for this reason that the distribution is also called the ageless lifespan distribution. The term “ageless” is explained in section 1.1.2.

Its average lifespan is the expected value of the random variable for the lifespan: $\frac{1}{\lambda}$.

Its half-life period is the instant $T = \ln(2)/\lambda$ with $\mathbb{P}(X > T) = 1/2$.

1.1.2. *The properties of an exponential distribution*

1.1.2.1. *The memoryless distribution*

PROPOSITION 1.1.– *The exponential distribution has no memory.*

In games of chance, the number of attempts needed to win does not depend on the number of attempts already made. This property is called *the memoryless property* or *the amnesia property*.

The exponential distribution also satisfies this property, which is expressed by:

$$\mathbb{P}(X > s + t \mid X > s) = \mathbb{P}(X > t), \qquad \forall s, t \in \mathbb{R}^+ \tag{1.5}$$

PROOF.–

By using Bayes'[2] theorem for any random variable X, we can write for all positive, real numbers s and t:

$$\mathbb{P}(X > s + t \mid X > s) = \frac{\mathbb{P}(X > s + t \ and \ X > s)}{\mathbb{P}(X > s)} = \frac{\mathbb{P}(X > s + t\,)}{\mathbb{P}(X > s)} = \frac{F_X^C(s + t)}{F_X^C(s)}$$

where $F_X^C(t) = \mathbb{P}(X > t)$ designates the complementary cumulative distribution function for the random variable X.

Equation [1.5] can thus be expressed as:

$$F_X^C(s + t) = F_X^C(s)F_X^C(t), \qquad \forall s, t \in \mathbb{R}^+ \tag{1.6}$$

Since F_X^C is a complementary cumulative distribution function, it is monotone and bounded, and given that $F_X^C(t) \neq 1 \ \forall t \in \mathbb{R}^+$, then F_X^C is necessarily an exponential function. The constant k therefore exists such that:

$$F_X^C(t) = e^{kt}, \qquad \forall t \in \mathbb{R}^+ \tag{1.7}$$

Now, $F_X^C < 1$, so k must be a negative constant, which we denote by $k = -\lambda$, $\lambda > 0$. We find that the complementary cumulative distribution function of a real, random variable follows the exponential distribution of the parameter λ. ■

If the lifespan X of an object follows an exponential distribution, the probability that this object will last at least $s + t$ units of time, keeping in mind that it has already lasted s units of time, is the same as the probability of lasting t units of time from its birth. In this case, we say that its lifespan is *memoryless* or *ageless*.

1.1.2.2. *The minimum of independent exponential distributions*

Let there be n independent random variables $X_1, X_2, \ldots$ and X_n following exponential distributions with the respective parameters $\lambda_1, \lambda_2, \ldots$ and λ_n. We denote the sum of these parameters with λ.

PROPOSITION 1.2.– *The random variable X, equal to the minimum of* $X_1, X_2, \ldots, X_n$, *follows an exponential distribution with parameter* λ.

2 Thomas Bayes (1702–1761), British mathematician and pastor of a Presbyterian Church.

PROPOSITION 1.3.– *The random variable X is equal to X_i ($i \in \{1, 2, \dots, n\}$) with a probability λ_i/λ, independently of the value of X.*

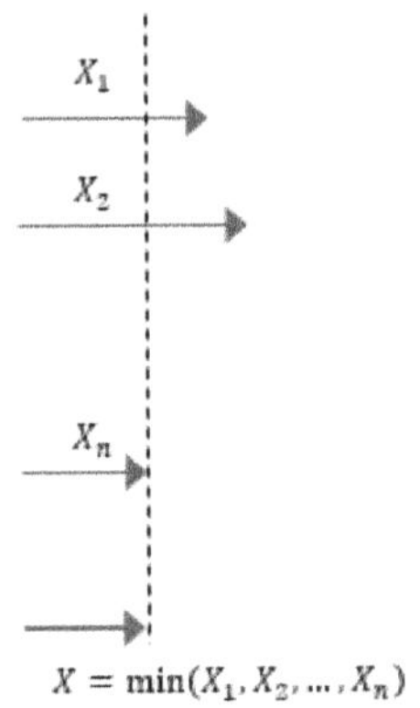

Figure 1.1. *Minimum of exponential distributions*

PROOF.–

The minimum $X = \min(X_1, X_2, \dots, X_n)$ of these random variables satisfies for any real number $t > 0$ (keeping in mind that the random variables X_i are independent of each other):

$$\begin{aligned}\mathbb{P}(X > t) &= \mathbb{P}(X_1 > t, X_2 \geq t, \dots, X_n \geq t) \\ &= \mathbb{P}(X_1 > t).\mathbb{P}(X_2 \geq t) \dots \mathbb{P}(X_n > t) \\ &= e^{-\lambda_1 t}.e^{-\lambda_2 t} \dots e^{-\lambda_n t} = e^{-(\lambda_1+\lambda_2+\cdots+\lambda_n)t} \\ &= e^{-\lambda t}\end{aligned}$$

For X to be equal to X_i:

$$\mathbb{P}(X > t, X = X_i) = \mathbb{P}(X_i > t, X_1 \geq X_i, X_2 \geq X_i, \dots, X_n \geq X_i)$$

$$\begin{aligned}&= \int_t^{+\infty} \lambda_i e^{-\lambda_i s}.e^{-\lambda_1 s} e^{-\lambda_2 s} \dots e^{-\lambda_n s}\, ds \\ &= \int_t^{+\infty} \lambda_i e^{-(\lambda_1+\lambda_2+\cdots+\lambda_i+\cdots+\lambda_n)s}\, ds \\ &= \frac{\lambda_i e^{-(\lambda_1+\lambda_2+\cdots+\lambda_n)t}}{\lambda_1+\lambda_2+\cdots+\lambda_n} = \frac{\lambda_i}{\lambda} e^{-\lambda t} \qquad \blacksquare\end{aligned}$$

The properties in Propositions 1.2 and 1.3 also apply in the case of an infinite family $\{X_n\}_{n\geq 1}$ of random variables following exponential distributions with the respective parameters from the infinite sequence $\{\lambda_n\}_{n\geq 1}$ when the sum of these parameters is finite:

$$\lambda = \sum_{n=1}^{+\infty} \lambda_n < \infty \qquad [1.8]$$

PROPOSITION 1.2b.– *The random variable* X*, equal to the minimum of* $X_1, X_2, \ldots$*, follows an exponential distribution with parameter* λ.

PROPOSITION 1.3b.– *The random variable* X *is equal to* X_i *(*$i \in \mathbb{N}^*$*) with a probability* λ_i/λ*, independently of the value of* X.

1.1.2.3. *The sum of independent exponential distributions*

Let $\{X_n\}_{n\geq 1}$ be a sequence of independent and identically exponentially distributed random variables with parameter λ. We call the sum of the first *n* terms of this sequence $S_n = X_1 + X_2 + \cdots + X_n$. We posit that $S_0 = 0$.

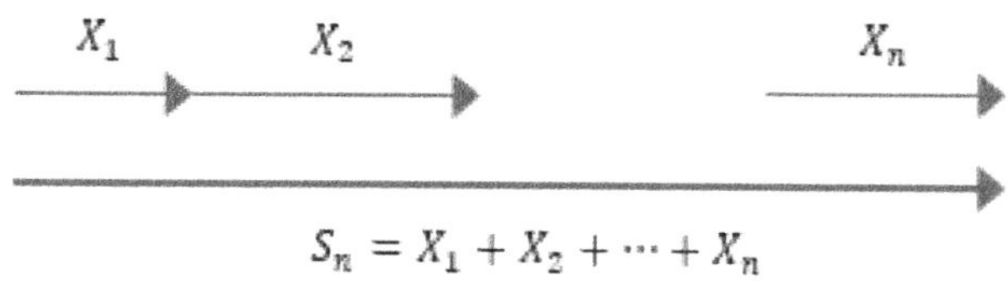

Figure 1.2. *Sum of exponential distributions*

PROPOSITION 1.4.– *The random variable* $S_n = X_1 + X_2 + \cdots + X_n$ *follows an Erlang*[3] *distribution with parameters (n,* λ*).*

PROOF.–

We call $N(t)$ *the integer-valued random variable equal to* n *satisfying* $S_n \leq t \leq S_{n+1}$ *for any real positive t. We first show that* $N(t)$ *follows the Poisson distribution with parameter* λt*:*

$$\mathbb{P}(N(t) = n) = \mathbb{P}(S_n \leq t \leq S_{n+1})$$

3 Agner Krarup Erlang (1878–1929), Danish mathematician.

$$= \int_{t_1+\cdots+t_n \leq t} (\lambda e^{-\lambda t_1})(\lambda e^{-\lambda t_2}) \ldots (\lambda e^{-\lambda t_n}).\, e^{-\lambda(t-t_1-\cdots-t_n)}\, dt_1 \ldots dt_n$$

$$= \int_{t_1+\cdots+t_n \leq t} \lambda^n e^{-\lambda(t_1+t_2+\cdots+t_n)} e^{-\lambda(t-t_1-\cdots-t_n)}\, dt_1 \ldots dt_n$$

$$= \lambda^n e^{-\lambda t} \int_{t_1+\cdots+t_n \leq t} dt_1 \ldots dt_n = \frac{(\lambda t)^n}{n!} e^{-\lambda t}$$

From this distribution, we can deduce the complementary cumulative distribution function $\mathbb{P}(S_n > t)$ for the random variable S_n:

$$\mathbb{P}(S_n > t) = \sum_{k=0}^{n-1} \mathbb{P}(N(t) = k) = e^{-\lambda t}\left(1 + \lambda t + \cdots + \frac{(\lambda t)^{n-1}}{(n-1)!}\right)$$

$$= e^{-\lambda t} \sum_{k=0}^{n-1} \frac{(\lambda t)^k}{k!}$$

Thus, the sum S_n follows an Erlang distribution with parameters (n, λ), which we denote by $\mathcal{E}(n, \lambda)$. ■

REMARKS.–

The Erlang distribution is a gamma distribution whose shape parameters are integers.

The density function of the Erlang distribution $\mathcal{E}(n, \lambda)$ is equal to:

$$f(x) = \frac{\lambda^n x^{n-1}}{(n-1)!} e^{-\lambda x} \qquad [1.9]$$

1.1.2.4. *The geometric sum of exponential distributions*

Let $\{X_n\}_{n\geq 1}$ be a sequence of i.i.d random variables following an exponential distribution with parameter λ. We denote by $S = X_1 + X_2 + \cdots + X_N$ the sum of the first N terms of this sequence such that N is an integer-valued random variable following a geometric distribution with parameter $p \in]0,1]$.

Thus, the random variable S is equal to:

– X_1 with a probability p;

– $X_1 + X_2$ with a probability $p(1-p)$;

– $X_1 + X_2 + \cdots + X_N$ with a probability $p(1-p)^{N-1}$ for any integer $N \in \mathbb{N}$.

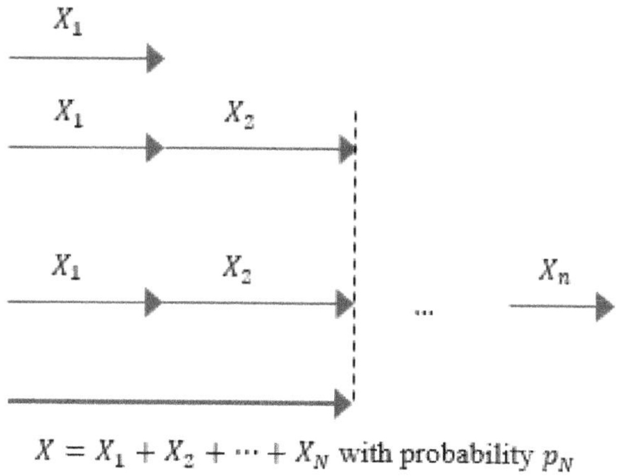

Figure 1.3. *Geometric sum of exponential distributions*

PROPOSITION 1.5.– *The random variable S follows an exponential distribution with parameter pλ.*

PROOF.–

We determine the complementary cumulative distribution function of the random variable S while simultaneously considering the random variable S_n from Proposition 1.4:

$$\mathbb{P}(S > t) = \sum_{n=1}^{+\infty} p(1-p)^{n-1}\mathbb{P}(S_n > t)$$

$$= \sum_{n=1}^{+\infty} p(1-p)^{n-1} e^{-\lambda t} \sum_{k=0}^{n-1} \frac{(\lambda t)^k}{k!}$$

$$= \sum_{n=1}^{+\infty} p(1-p)^{n-1} e^{-\lambda t} \sum_{k=0}^{n-1} \frac{(\lambda t)^k}{k!}$$

$$= \sum_{k=0}^{+\infty} e^{-\lambda t} \frac{(\lambda t)^k}{k!} \sum_{n>k} p(1-p)^{n-1}$$

$$= \sum_{k=0}^{+\infty} e^{-\lambda t} \frac{(\lambda t)^k}{k!} (1-p)^k$$

$$= \sum_{k=0}^{+\infty} e^{-\lambda t} \frac{((1-p)\lambda t)^k}{k!}$$
$$= e^{-p\lambda t}, \quad \forall t \in \mathbb{R}^+ \qquad \blacksquare$$

1.2. Poisson process

1.2.1. *Definitions*

DEFINITION 1.2.– *A random process (or a stochastic process) is a representation of the evolution over time of a random variable.*

A random process is therefore a sequence of indexed random variables $X(t) \in \mathcal{X}$ over time $t \in \mathcal{T}$. We examine a particular case of the random process for $\mathcal{T} = \mathbb{N}^* = \{1,2,\dots\}$: it is the punctual process for which we denote $X(t) = T_t$ for all $t \in \mathcal{T}$.

DEFINITION 1.3.– *A punctual process over* $\mathbb{R}^+$ *is a growing sequence* $T_1, T_2, \dots$ *of real positive random variables typically representing instants when events occurs.*

EXAMPLE.– The arrival of customers in a queue is a punctual process. The instants of their arrivals, which are random variables, form a growing sequence.

The random variables $\mathrm{T}_1, \mathrm{T}_2, \dots$ are called *the instants of appearance of the process*.

We denote by τ_n $(n \geq 1)$ the inter-arrival times of the process: $\tau_n = T_n - T_{n-1}$, with $T_0 = 0$.

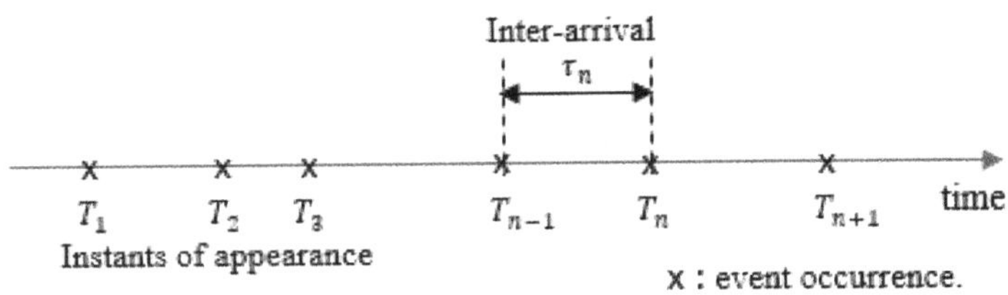

Figure 1.4. *Punctual process*

DEFINITION 1.4.– *A punctual process is called simple if the inter-arrivals are almost surely (AS) strictly positive.*

DEFINITION 1.5.– *A punctual process is called stationary if, for any* $n \geq 0$*, the random variables of inter-arrivals* $\{\tau_n\}_{n\geq 1}$ *follow the same distribution.*

In the case of a stationary punctual process, the inter-arrivals therefore have the same expected value: the average of the inter-arrivals. The inverse of this average is called the *intensity of the process,* denoted λ:

$$\lambda = \frac{1}{E(\tau_n)} = \frac{1}{E(\tau_1)} \qquad [1.10]$$

DEFINITION 1.6.– *When the times of inter-arrivals form a sequence of independent random variables following exponential distributions with parameter* $\lambda > 0$*, we say that the process is a Poisson process of intensity* λ*.*

PROPOSITION 1.6.– *A Poisson process is a simple and stationary punctual process.*

PROOF.–

This property derives directly from the definition of the Poisson process. ■

The counting measure N of a punctual process counts the number of appearances (or the number of arrivals) of the process in an interval of time (s,t):

$$N(s,t) = \sum_{n=1}^{+\infty} 1_{(s,t)}(T_n), \quad \forall s,t \in \mathbb{R}^+, s < t \qquad [1.11]$$

where $1_{(s,t)}(x)$ indicates the indicator function of the interval (s,t) at point x: it equals 1 if $x \in (s,t)$ and 0 if $x \notin (s,t)$.

We can also use the notation $N(t)$ to count the number of appearances of the process up to the instant t: $N(t) = N(0,t)$.

This counting measure also allows us to describe the punctual process:

– the process is simple if the probability that $N(s,t) \leq 1$ given that $N(s,t) \geq 1$ tends towards 1 when s tends towards t;

– the process is stationary if the random variable $N(s,t)$ has the same distribution as $N(0,t-s)$ for any interval (s,t). The intensity of the process is therefore given by the equation:

$$\lambda = E(N(0,1)) \qquad [1.12]$$

– a Poisson process of intensity $\lambda > 0$ is a punctual process such that the numbers of appearances of the process for all discontinuous intervals are independent and the number of appearances of the process for any interval (s,t) follows a Poisson distribution with parameter $\lambda(t-s)$:

$$\mathbb{P}(N(s,t)=n) = \frac{\left(\lambda(t-s)\right)^n}{n!} e^{-\lambda(t-s)}, \quad \forall n \in \mathbb{N} \qquad [1.13]$$

REMARKS.–

From equation [1.12], we can say that the intensity λ of the process corresponds to the average number of appearances of the process per unit of time. This is why we also call it the arrival rate of the process.

Following from the stationariness of any process, the number of appearances in the intervals $(0,1),(1,2),\ldots,(n,n+1),\ldots$ are i.i.d random variables with the average λ. From the law of large numbers, we obtain:

$$\lim_{t\to\infty} \frac{1}{t} N(t) = E\left(N(0,1)\right) = \lambda \quad A.S.$$

From Proposition 1.4, the instant of the nth appearance of the Poisson process of intensity λ is a random variable following the Erlang distribution $\mathcal{E}(n,\lambda)$.

1.2.2. *Properties of the Poisson process*

1.2.2.1. *Memoryless process*

PROPOSITION 1.7.– *A Poisson process is memoryless.*

Saying that the Poisson process is memoryless means that knowledge about the process up to time t says nothing about its future evolution.

PROOF.–

We denote as $N(t) = n$ the number of appearances of the process up to time t, and we determine the probability of k appearances in an interval $(t, s+t)$:

$$\mathbb{P}(N(t,s+t)=k|N(t)=n) = \frac{\mathbb{P}(N(t,s+t)=k \ \ and \ \ N(t)=n)}{\mathbb{P}(N(t)=n)}$$

Independently of $N(t, s+t)$ and $N(t)$, we have:

$$\mathbb{P}(N(t,s+t)=k|N(t)=n)=\frac{\mathbb{P}(N(t,s+t)=k).\mathbb{P}(N(t)=n)}{\mathbb{P}(N(t)=n)}$$

Through the stationariness of the Poisson process, $\mathbb{P}(N(t,s+t)=k)=\mathbb{P}(N(s)=k)$.

Thus,

$$\mathbb{P}(N(t,s+t)=k|N(t)=n)=\mathbb{P}(N(s)=k)=\frac{(\lambda s)^k}{k!}e^{-\lambda s}$$

The obtained probability depends on neither the time t nor the number n of appearances of the process up to time t. Knowledge of the process up to time t says nothing about its future evolution; hence, its amnesia property.■

REMARK.– The memoryless property distinguishes the Poisson process among simple stationary punctual processes.

1.2.2.2. *Distribution of points in a Poisson process*

PROPOSITION 1.8.– *Given that an interval (s,t) of $\mathbb{R}^+$ contains n points of a Poisson process, these points are uniformly distributed in this interval.*

PROOF.–

For any integer $l \geq 1$, any division of the interval (s,t) in l intervals $(s,t_1),(t_1,t_2),\dots,(t_{l-1},t)$ and all integers $n_1,n_2,\dots,n_l$ with a sum equal to n, we have:

$$\begin{aligned}\mathbb{P}(N(s,t_1)&=n_1,\dots,N(t_{l-1},t)=n_l \mid N(s,t)=n)\\&=\frac{\mathbb{P}(N(s,t_1)=n_1,\dots,N(t_{l-1},t)=n_l)}{\mathbb{P}(N(s,t)=n)}\end{aligned}$$

$$=\frac{e^{-\lambda(t_1-s)}\frac{\left(\lambda(t_1-s)\right)^{n_1}}{n_1!}\dots e^{-\lambda(t-t_{l-1})}\frac{\left(\lambda(t-t_{l-1})\right)^{n_l}}{n_l!}}{e^{-\lambda(t-s)}\frac{\left(\lambda(t-s)\right)^{n}}{n!}}$$

$$=\binom{n}{n_1,\dots,n_l}\left(\frac{t_1-s}{t-s}\right)^{n_1}\dots\left(\frac{t-t_{l-1}}{t-s}\right)^{n_l}$$

The probability that any one of the l intervals contains one of the n points of the Poisson process is therefore proportional to its length: each point is distributed uniformly across the interval (s,t), independently of the other points.■

1.2.2.3. *Superposition of the Poisson process*

PROPOSITION 1.9. (Palm[4] theorem).– *The superposition of K independent Poisson processes of intensity $\lambda_1, \dots, \lambda_K$ is a Poisson process of intensity $\lambda = \lambda_1 + \cdots + \lambda_K$.*

PROOF.–

We denote by $N_1, N_2, \dots, N_K$ the counting measures of K initial Poisson processes, and we determine the counting measure N of the superposed process:

$$\mathbb{P}(N(s,t) = n) = \sum_{\substack{n_1,\dots,n_K \\ n_1+\cdots+n_K=n}} \mathbb{P}(N_1(s,t) = n_1, \dots, N_K(s,t) = n_K)$$

The initial Poisson processes are independent. Since a Poisson process is simple, the initial Poisson processes AS have nothing in common. Thus, we can write:

$$\mathbb{P}(N(s,t) = n) = \sum_{\substack{n_1,\dots,n_K \\ n_1+\cdots+n_K=n}} \mathbb{P}(N_1(s,t) = n_1) \dots \mathbb{P}(N_K(s,t) = n_K)$$

$$= \sum_{\substack{n_1,\dots,n_K \\ n_1+\cdots+n_K=n}} \mathbb{P}(N_1(s,t) = n_1) \dots \mathbb{P}(N_K(s,t) = n_K)$$

$$= \sum_{\substack{n_1,\dots,n_K \\ n_1+\cdots+n_K=n}} \frac{\left(\lambda_1(t-s)\right)^{n_1}}{n_1!} e^{-\lambda_1(t-s)} \dots \frac{\left(\lambda_K(t-s)\right)^{n_K}}{n_K!} e^{-\lambda_K(t-s)}$$

$$= \sum_{\substack{n_1,\dots,n_K \\ n_1+\cdots+n_K=n}} \binom{n}{n_1, \dots, n_K} \left(\lambda_1(t-s)\right)^{n_1} \dots \left(\lambda_K(t-s)\right)^{n_K} e^{-\lambda(t-s)}$$

$$= \frac{\left(\lambda(t-s)\right)^n}{n!} e^{-\lambda(t-s)}$$

We thus find the counting measure of a Poisson process of intensity λ. ■

4 Conny Palm de Conrad (1907–1951), Swedish engineer, electrician and statistician.

Figure 1.5 is a representation of this superposition of two Poisson processes into one process.

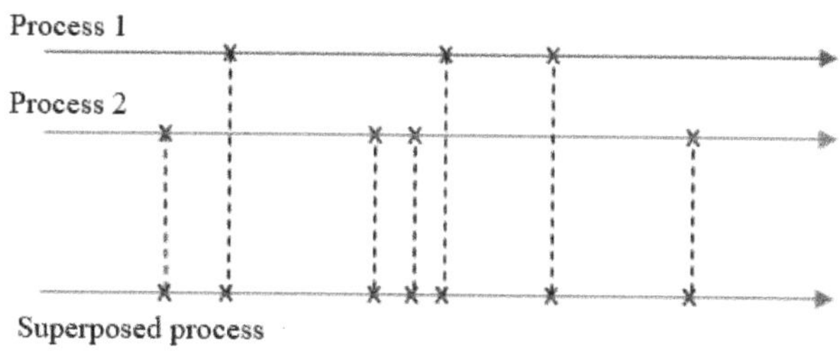

Figure 1.5. *Superposition of Poisson processes*

PROPOSITION 1.10.– *The probability that any one point of a superposed process belongs to one of the K initial processes is proportional to the intensity of this process.*

PROOF.–

$$\begin{aligned}
&\mathbb{P}(N_i(s,t) = 1 \mid N(s,t) = 1) \\
&= \frac{\mathbb{P}(N_i(s,t) = 1, N_1(s,t) = 0, \dots, N_K(s,t) = 0)}{\mathbb{P}(N(s,t) = 1)} \\
&= \frac{\lambda_i (t-s) e^{-\lambda(t-s)}}{\lambda(t-s)e^{-\lambda(t-s)}} = \frac{\lambda_i}{\lambda} \quad \blacksquare
\end{aligned} \qquad [1.14]$$

1.2.2.4. *Subdivision of a Poisson process*

Let N be the counting measure of a Poisson process of intensity λ and $p_1, \dots, p_K$ K strictly positive real numbers with a sum equal to 1.

PROPOSITION 1.11. (Raikov[5] theorem).– *The K punctual processes obtained by attributing each point of the Poisson process to the* kth *punctual process of probability* p_k *are independent Poisson processes of respective intensities* $\lambda p_1, \dots, \lambda p_K$.

5 Dmitrii Abramovich Raikov (1905), Russian professor with a PhD in mathematical physics.

PROOF.–

For any interval (s,t), the counting measures $N_1, \ldots, N_K$ of K obtained punctual processes satisfy:

$$\begin{aligned}\mathbb{P}(N_1(s,t) = n_1, \ldots, N_K(s,t) = n_K) \\ = \binom{n}{n_1, \ldots, n_K} p_1^{n_1} \ldots p_K^{n_K} . \frac{\lambda^n (t-s)^n}{n!} e^{-\lambda(t-s)} \\ = \frac{(\lambda p_1 (t-s))^{n_1}}{n_1!} e^{-\lambda p_1 (t-s)} \ldots \frac{(\lambda p_K (t-s))^{n_K}}{n_K!} e^{-\lambda p_K (t-s)}\end{aligned} \qquad [1.15]$$

with $n = n_1 + \cdots + n_K$.■

Figure 1.6 represents this subdivision of the Poisson process.

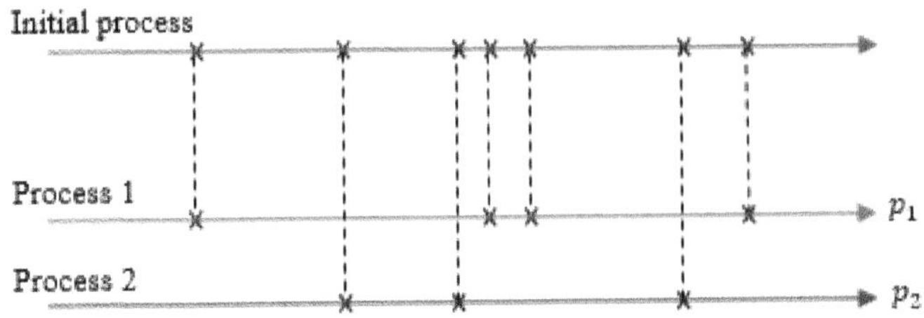

Figure 1.6. *Subdivision of a Poisson process*

1.3. Exercises

EXERCISE 1.1.– *Radioactive particle*

The lifetime of a radioactive particle follows an exponential distribution with parameter λ.

1) What is its lifetime?

2) Calculate the probability that the particle will still exist after its expected lifetime.

3) What is its expected lifetime given that the particle still exists after its expected lifetime?

EXERCISE 1.2.– *Discretization of exponential distribution*

Let X be a random exponential variable with parameter λ.

1) Show that the ceiling function of X follows a geometric distribution which we can determine.

2) Show that the floor function of X follows a geometric distribution which we can determine.

3) More generally, what is the distribution of an integer-valued random variable's ceiling function of X/τ for any real number > 0?

EXERCISE 1.3.– *Synchronization*

Let X_1 and X_2 be two independent exponentially distributed random variables with respective parameters λ_1 and λ_2.

Show that the variables X_1 and X_2 are equal with a probability of zero.

Directions: we can use $\mathbb{P}(A_1 \cap A_2) + \mathbb{P}(A_1 \cup A_2) = \mathbb{P}(A_1) + \mathbb{P}(A_2)$ for all events A_1 and A_2.

EXERCISE 1.4.– *"Very" random variable*

Let X be a real positive random variable with a probability density function $f_X(x)$. The entropy of the random variable X is equal to:

$$h(X) = -\int_0^{+\infty} f_X(t).\ln\big(f_X(t)\big)\,dt$$

1) Calculate the entropy of the random variable X_e following an exponential distribution with parameter λ.

2) Using the Gibbs[6] inequality for all probability density functions $f(x)$ and $g(x)$ over $\mathbb{R}^+$: $\int_0^{+\infty} f(t).\ln\big(f(t)/g(t)\big)\,dt \geq 0$, with an inequality if, and only if, $f = g$, show that, for any random variable X with a probability density $f_X(x)$ and an average $1/\lambda$, we get:

$$h(X) \leq 1 - \ln \lambda$$

and that we have inequality if, and only if, X is a random variable following an exponential distribution with parameter λ.

6 Willard Gibbs (1839–1903), American physicist and mathematician, professor of mathematical physics.

3) Deduce from this that the exponential distribution is the distribution of maximum entropy among all distributions with the same average and density over $\mathbb{R}^+$.

EXERCISE 1.5.– *Call expiry*

A personal call and a professional call have durations following exponential distributions with respective averages of 30 seconds and one minute.

Supposing these two calls (a personal call and a professional call) begin at the same instant:

1) After how long can we expect a call to be completed?

2) What is the probability that the professional call will finish first?

Supposing that, at any instant t, we count n personal calls and m professional calls taking place at the same time:

3) What is the probability that a professional call will finish first?

EXERCISE 1.6.– *Inter-arrivals*

In a call center, we suppose that the inter-arrivals of calls follow an exponential distribution with parameter λ. Office hours begin at 8 am.

1) Calculate the probability that no calls will come in during the first half-hour of working hours.

2) What is the probability that no calls will come in during the second half-hour of working hours, given that no calls came in during the first half-hour?

3) Calculate the probability that the second call will come in during the first half-hour of work.

4) Calculate the probability that the nth call will come in after instant t.

EXERCISE 1.7.– *ATM*

Customers approach an ATM following a Poisson process. The rate of arrival of customers is one every 15 minutes.

1) Calculate the probability that no customer will approach for one hour.

2) Calculate the probability that n successive customers will arrive in the interval of one minute.

3) If a customer arrives at the ATM, how much time has passed on average since the arrival of the previous customer?

EXERCISE 1.8.– *Bus*

To reach point B from point A, we can take one of two bus routes: line 1 or line 2. The buses for these lines arrive following Poisson processes of respective intensities λ_1 and λ_2 at stop A.

1) What is the average delay time for the next bus at stop A?

2) What is the probability that the bus for line 2 will arrive first at stop A?

3) If we prefer the bus for line 1, how long has passed since the last bus for this route passed?

Now, we only consider the buses for line 1.

4) What is the average delay time for the next bus?

5) From the perspective of the user showing up at stop A at an arbitrary instant, show that the average length of time between two buses is the double of the average effective length of time between two buses $1/\lambda_1$.

EXERCISE 1.9.– *Gas station*

In a gas station, the pumps are spread out in two different rows. At this station, we count an average of 15 cars arriving in 10 minutes; these arrivals are presumed to follow a Poisson process. We estimate that the car arriving uniformly and randomly picks which row it will join.

1) Calculate the probability that no cars will choose the first row in the station for one hour.

2) What is the average number of cars arriving in the second row in the space of one hour?

3) What is the probability that more than one minute will pass between two arrivals in the second row?

EXERCISE 1.10.– *Interlacing*

Let there be two Poisson processes of respective intensities λ_1 and λ_2.

Calculate the average number of appearances of one between two appearances of the other.

EXERCISE 1.11.– *Discretization of the Poisson process*

Let there be a growing sequence of random variables $\{t_n\}_{n\in\mathbb{N}^*}$ with integer values. In the case where the random variables $d_n = t_n - t_{n-1}$are i.i.d following geometric distributions with parameter $p \in (0,1)$, we speak of a punctual process over $\mathbb{N}$ called the Bernoulli[7] process.

For this process, the number of arrivals X_k at instant k, for any integer $k \geq 1$, follows a Bernoulli distribution with the average p defined by: $\mathbb{P}(X_k = 1) = p$, $\mathbb{P}(X_k = 0) = 1 - p$. The intensity of this simple and stationary process is equal to $p = E(X_i)$.

1) For any natural integer $s < t$, calculate its counting measure in an interval (s, t), by analogy with the Poisson process.

2) Deduce that the Bernoulli process of intensity $p = \lambda\tau$ (fixed, strictly positive λ and a sufficiently small time step τ) tends towards a Poisson process of intensity λ when τ approaches zero.

7 Daniel Bernoulli (1700–1782), Swiss doctor, physicist and mathematician.

2

Markov Chains

If the present is born of the past, then the present is pregnant with the future.

Voltaire (1694–1778)

In this chapter, we review Markov[1] chains for discrete periods of time and Markov processes for continuous time. The previous chapter was limited to a series of independent events, but this time we observe the evolution of a system over time, supposedly correlated, satisfying the Markov property: its state at time $n + 1$ only depends on its state at time n. Given its present state, the future state of the system does not depend on its past states.

2.1. Markov chains in discrete time

2.1.1. *Definitions*

2.1.1.1. *Random process*

DEFINITION 2.1.– *A random process is the representation of the evolution of a random variable over a period of time.*

In other words, a random process is a series of random variables $\mathrm{X(t)} \in \mathcal{X}$ indexed by time $\mathrm{t} \in \mathcal{T}$. For a countable set $\mathcal{T}$, we say that the process is discrete.

The state $X(t)$ of a dynamic system evolving over a period of time in a random manner can be modeled by a random variable. This system is, informally, described by a random process.

1 Andrey Andreyevich Markov (1903–1979), Soviet mathematician, physicist and chemist.

DEFINITION 2.2.– *Informally, a random process is a dynamic system whose state changes over a period of time in a random manner.*

From now on, we consider $\mathcal{X} = \{0, \dots, n\} \subset \mathbb{N}$ *and* $\mathcal{T} = \mathbb{N}$, *that is, the possible state of this system is an integer between* 0 and n and that time t is a positive integer or zero.

We denote by $\pi_i(t)$ the probability that the system will be in state $X(t) = i$ at instant t, in other words $\pi_i(t) = \mathbb{P}(X(t) = i)$.

DEFINITION 2.3.– *A row vector of real numbers is called a stochastic vector if, and only if,*

– *all its coordinates are positive or zero;*

– *the sum of its coordinates is equal to* 1.

At any instant t, the vector $\boldsymbol{\pi}(t) = \big(\pi_0(t), \pi_1(t), \dots, \pi_n(t)\big)$ is a stochastic vector: $\pi_0(t) + \pi_1(t) + \dots + \pi_n(t) = 1 \ \forall t \in \mathbb{N}$.

A process can be defined by the application $t \mapsto \boldsymbol{\pi}(t)$, which designates the probabilities of states i at each instant t.

2.1.1.2. *Markov chains in discrete time*

DEFINITION 2.4.– *A Markov chain is a random process satisfying the Markov property: the key information for the prediction of the future* (*time* $\mathrm{t} + 1$) *is entirely contained in the present state* (*time* t) *of the process and does not depend on previous states* (*time before or equal to* $\mathrm{t} - 1$).

From Definition 2.4, we can deduce another definition of the Markov chain for the sequence of random variables $\{\mathrm{X(t)}\}_{\mathrm{t} \in \mathbb{N}}$ that we used in section 2.1.1.1.

DEFINITION 2.5. (Weak Markov property).– *A sequence of random variables* $\{X(t)\}_{t \in \mathbb{N}}$ *with values in a discrete set* $\mathcal{X}$ *is a Markov chain if it is memoryless in the sense that for any integer-valued positive time or zero, and for all states* $i, j, x_{t-1}, \dots, x_1, x_0 \in \mathcal{X}$:

$$\begin{aligned} &\mathbb{P}(X(t+1) = j \mid X(t) = i, X(t-1) = x_{t-1}, \dots, X(0) = x_0) \\ &\quad = \mathbb{P}(X(t+1) = j \mid X(t) = i) \end{aligned} \qquad [2.1]$$

DEFINITION 2.6.– *The Markov chain is called homogeneous over time, or simply homogeneous, if expression* [2.1] *does not depend on time* t.

For everything that follows, unless otherwise stated, we only consider Markov chains to be homogeneous.

Let there be two states i and j from the discrete set $\mathcal{X}$ of possible states of some system. The probability that this system will go from state i to state j is denoted by P_{ij} and only depends on states i and j. Such a system is called *memoryless*, or even *Markovian*, which means that the probability P_{ij} does not depend on the states that came before i through which the system has gone through in its history. Its future state only depends on its present state and not its past states.

DEFINITION 2.7.– *The probabilities of a transition P_{ij} set out in the matrix $\mathbf{P} = \left(P_{ij}\right)_{0\leq i,j\leq n}$ form what we call a transition matrix of the Markov chain.*

A Markov chain can therefore be defined by its transition probabilities or by its transition matrix:

$$\forall i,j \in \mathcal{X}, \qquad P_{ij} = \mathbb{P}(X_1 = j \mid X_0 = i) \qquad [2.2]$$

DEFINITION 2.8.– *A square matrix is called a stochastic matrix if each of its rows constitutes a stochastic vector.*

PROPOSITION 2.1.– *The transition matrix of a Markov chain is a stochastic matrix.*

PROOF.–

According to the definition of transition probability, we must have $\forall i \in \mathcal{X}, \sum_{j\in\mathcal{X}} P_{ij} = 1$: the vector in row i contains the probabilities of all possible transitions beginning with state i after which their sum is equal to one. All the rows in the transition matrix are thus stochastic vectors. ■

We can easily verify that all the powers of a stochastic matrix are also stochastic matrices.

The sequence of stochastic vectors $\boldsymbol{\pi}(t)$ for $t \in \mathbb{N}$ satisfies the matrix recurrence formula:

$$\boldsymbol{\pi}(t + 1) = \boldsymbol{\pi}(t).\mathbf{P}, \qquad \forall t \in \mathbb{N} \qquad [2.3]$$

that is:

$$\left(\pi_0(t+1), \pi_1(t+1), \dots, \pi_n(t+1)\right)$$

$$= \left(\pi_0(t), \pi_1(t), \dots, \pi_n(t)\right) \begin{pmatrix} P_{00} & P_{01} & \cdots & P_{0n} \\ P_{10} & P_{11} & & P_{1n} \\ \vdots & & \ddots & \vdots \\ P_{n0} & P_{n1} & \cdots & P_{nn} \end{pmatrix} \qquad [2.4]$$

DEFINITION 2.9.– *A homogeneous Markov chain over time is defined by a sequence of stochastic vectors* $\{\boldsymbol{\pi}(t)\}_{t \in \mathbb{N}}$ *and a stochastic matrix* **P** *such that for any* $t \in \mathbb{N}$, $\boldsymbol{\pi}(t+1) = \boldsymbol{\pi}(t)\mathbf{P}$.

The chain is homogeneous over time because the matrix **P** does not depend on time t.

DEFINITION 2.10.– *We call the directed graph whose vertices are coordinates of* $\mathcal{X}$ *and whose arcs represent the non-zero probability transitions the Laplace transform, as illustrated in Figure 2.1.*

Each arc is weighted with the probability of the corresponding transition. The chain is called *irreducible* if its transition graph is strongly connected, that is, if for all distinct states $i, j \in \mathcal{X}$, a path exists in the graph from i to j and a path from j to i (see section 2.1.2.3).

From here on, we only consider irreducible Markov chains unless otherwise stated.

EXAMPLE (Two unrepairable machines).– Let there be a machine containing two components that function independently of each other. Each component has a reliability equal to p over the course of a day, which means that the probability of it breaking down during this period is $1 - p$. There is no way of repairing it. At first, the two components of the machine function correctly.

We consider the state of the machine to be the number of components that are out of order at the beginning of the day. The possible states are 0, 1 or 2 depending on whether there are 0, 1 or 2 components out of order at the start of the day. We thus have a random process such that $\mathcal{X} = \{0, 1, 2\}$. This process is a Markov chain since the number of components that are out of order at the start of day $t+1$ only depends on the number of components that are out of order at the start of day t, and not on that of the days preceding t.

The transition matrix of this Markov chain is equal to:

$$\mathbf{P} = \begin{pmatrix} P_{00} & P_{01} & P_{02} \\ P_{10} & P_{11} & P_{12} \\ P_{20} & P_{21} & P_{22} \end{pmatrix} = \begin{pmatrix} p^2 & 2p(1-p) & (1-p)^2 \\ 0 & p & 1-p \\ 0 & 0 & 1 \end{pmatrix}$$

The transitions between the states of the chain and their probabilities are represented by the arcs and their weight, respectively, in the graph in Figure 2.1.

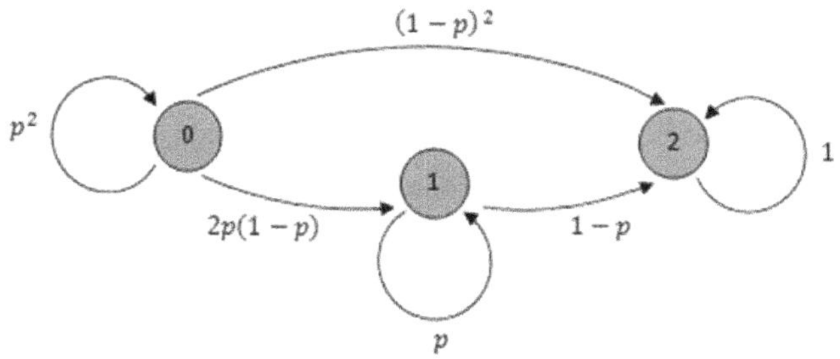

Figure 2.1. *Transition graph*

2.1.1.3. *Chapman–Kolmogorov equation*

Let there be times $s,\ t \in \mathbb{N}$ such that $s < t$.

We denote by $P_{ij}(s,t) = \mathbb{P}(X(t) = j \,|X(s) = i)$.

PROPOSITION 2.2 (Chapman[2]–Kolmogorov[3] equation).– *The probabilities of transitions between any two states i and j satisfy the following equation:*

$$P_{ij}(s,t) = \sum_{k=0}^{n} P_{ik}(s,r)P_{kj}(r,t) \qquad [2.5]$$

PROOF.–

We have:

$$P_{ij}(s,t) = \mathbb{P}(X(t) = j \,|X(s) = i) = \frac{\mathbb{P}(X(t) = j\,, X(s) = i)}{\mathbb{P}(X(s) = i\,)}$$

2 Sydney Chapman (1888–1970), British mathematician, astronomer and geophysicist.

3 Andrey Nikolaevich Kolmogorov (1903–1987), Soviet Russian mathematician.

$$= \frac{1}{\mathbb{P}(X(s) = i\,)} \sum_{k=0}^{n} \mathbb{P}(X(t) = j\,, X(r) = k, X(s) = i)$$

$$= \frac{1}{\mathbb{P}(X(s) = i\,)} \sum_{k=0}^{n} \mathbb{P}(X(t) = j\,|X(r) = k). \mathbb{P}(X(r) = k\,|X(s) = i)\,. \mathbb{P}(X(s) = i)$$

$$= \sum_{k=0}^{n} \mathbb{P}(X(t) = j\,|X(r) = k). \mathbb{P}(X(r) = k\,|X(s) = i)$$

$$= \sum_{k=0}^{n} P_{kj}(r,t)P_{ik}(s,r) = \sum_{k=0}^{n} P_{ik}(s,r)P_{kj}(r,t) \quad \blacksquare$$

More specifically, for a homogeneous Markov chain, we have, when using $\mathrm{P}_{\mathrm{ij}}^{(\mathrm{t})} = P_{ij}(0,t)$:

$$P_{ij}^{(t+1)} = \sum_{k=0}^{n} P_{ik}P_{kj}^{(t)} = \sum_{k=0}^{n} P_{ik}^{(t)}p_{kj} \qquad [2.6]$$

Note that according to the definition of the matrix product, $P_{ij}^{(t+1)}$ *is the coordinate of* $\mathbf{P}^{t+1}$.

2.1.2. *Evolution of a stochastic vector over time*

2.1.2.1. *State equations*

PROPOSITION 2.3.– *At any instant* t *(integer), the stochastic vector is expressed as:*

$$\boldsymbol{\pi}(t) = \boldsymbol{\pi}(0). \mathbf{P}^{t} \qquad [2.7]$$

PROOF.–

From the recurrence formula $\boldsymbol{\pi}(t+1) = \boldsymbol{\pi}(t).\mathbf{P}$ of equation [2.3] applied to instants $t = 0, 1, 2, ...$, we obtain the following expressions:

– $\boldsymbol{\pi}(1) = \boldsymbol{\pi}(0).\mathbf{P}$;

– $\boldsymbol{\pi}(2) = \boldsymbol{\pi}(1).\mathbf{P} = \boldsymbol{\pi}(0).\mathbf{P}^2$;

– $\boldsymbol{\pi}(3) = \boldsymbol{\pi}(2).\mathbf{P} = \boldsymbol{\pi}(0).\mathbf{P}^3$;

– Supposing that $\boldsymbol{\pi}(t) = \boldsymbol{\pi}(0).\mathbf{P}^t$ is true up until time t, we still have $\pi(t+1) = \pi(t).\mathbf{P} = \pi(0).\mathbf{P}^t.\mathbf{P} = \boldsymbol{\pi}(0).\mathbf{P}^{t+1}$ at time $t+1$.■

2.1.2.2. *Calculating the power of a matrix*

The calculation of the power of a matrix can be done iteratively through the formula $\mathbf{P}^t = \mathbf{P}.\mathbf{P}^{t-1}$ for $t \geq 1$.

In order to go faster, we can also use the equation $\mathbf{P}^t = (\mathbf{P}^2)^{t/2}$ when t is an even integer. With every pass through the calculation cycle, the value of the exponent t is divided by two. When t has an uneven value, we return to an even number by decreasing t.

We can also use the eigendecomposition of the matrix $\mathbf{P}$ in $\mathbf{P} = \mathbf{QDQ}^{-1}$, where $\mathbf{Q}$ is a matrix whose i-th column is the eigenvector $\mathbf{q}_i$ of $\mathbf{P}$ and $\mathbf{D}$ is the diagonal matrix whose diagonal coefficients are the eigenvalues λ_i: $\mathbf{D} = \text{diag}(\lambda_0, \lambda_1, \ldots, \lambda_n)$.

Recall that eigenvalues are the solutions to the equation $\det(\mathbf{P} - \lambda\mathbf{I}) = 0$, and eigenvectors are the solutions to the equation for each eigenvalue λ_i: $(\mathbf{P} - \lambda_i\mathbf{I}).\mathbf{q}_i = 0$.

In this case, we have $\mathbf{P}^t = \mathbf{QD}^t\mathbf{Q}^{-1}$ with $\mathbf{D}^t = \text{diag}(\lambda_0^t, \lambda_1^t, \ldots, \lambda_n^t)$.

2.1.2.3. *Ergodic chains*

DEFINITION 2.11.– *A state $j \in \mathcal{X}$ can be reached from another state $i \in \mathcal{X}$, and we say $i \rightsquigarrow j$, if, starting from i, we reach j with a positive probability in a finite number of steps:*

$$\exists t \in \mathcal{T}, \qquad p_{ij}^{(t)} > 0 \qquad [2.8]$$

where $p_{ij}^{(t)}$ designates the coordinate ij of $\mathbf{P}^t$.

We say $i \leftrightsquigarrow j$ if we have both $i \rightsquigarrow j$ and $j \rightsquigarrow i$ simultaneously.

DEFINITION 2.12.– *A Markov chain is called* irreducible *or* ergodic *if every state can be reached in a finite number of steps starting from any other state:*

$$\forall i, j \in \mathcal{X} : i \leftrightsquigarrow j \qquad [2.9]$$

From Definitions 2.11 and 2.12 we can establish the following proposition:

PROPOSITION 2.4.– *If a certain power of the transition matrix* $\mathbf{P}$ *only has strictly positive coordinates, then the chain is ergodic.*

PROOF.–

The proof is obvious from these definitions. ■

We speak of a *regular chain* when a certain power n exists for the transition matrix such that all the coordinates of $\mathbf{P}^n$ are strictly positive. Then, a regular chain is necessarily ergodic.

2.1.2.4. *Distribution limit*

We observe that the distribution $\boldsymbol{\pi}(t)$ often converges to a distribution limit denoted by $\boldsymbol{\pi}(\infty)$ when t tends towards infinity. For that, a principal property of regular (and therefore ergodic) chains is as follows:

PROPOSITION 2.5.– *Let* $\mathbf{P}$ *be the transition matrix of a regular chain, so a stochastic vector* $\boldsymbol{\pi}$ *exists that does not depend on* $\boldsymbol{\pi}(0)$ *such that the limit* $\lim_{t\to\infty} \boldsymbol{\pi}(t) = \boldsymbol{\pi}$. *Moreover,* $\boldsymbol{\pi}$ *satisfies the equation* $\boldsymbol{\pi} = \boldsymbol{\pi}\mathbf{P}$.

PROOF.–

If the chain only has one state, the result is immediate. In the rest of the demonstration, we suppose the chain has $N \geq 2$ states. We can suppose that all the coordinates of $\mathbf{P}^t$ *are strictly positive for* $t = 1$*; otherwise, it is sufficient to consider the chain whose transition matrix is* $\mathbf{P}^t$ *rather than* $\mathbf{P}$.

Let $d > 0$ be the smallest coordinate of $\mathbf{P}$. Thus, we have $d \geq 1/2$ since $Nd \leq 1$. Let $\mathbf{y}$ be a column vector such that all its coordinates are included between m_0 and M_0. Let $\mathbf{z} = \mathbf{Py}$. The greatest possible value of a coordinate z_j of $\mathbf{z}$ is obtained if $\mathbf{y}^T = (m_0, M_0, \ldots, M_0)$ and $p_{k_1} = d$. In this case, the sum of the $N-1$ last coordinates of row j of $\mathbf{P}$ is equal to $1-d$, and consequently $z_j = dm_0 + (1-d)M_0$.

We therefore necessarily have:

$$z_j \leq dm_0 + (1-d)M_0 := M_1 \qquad \forall j \in \{1, \ldots, N\}$$

Similar reasoning shows that:

$$z_j \geq dM_0 + (1-d)m_0 := m_1 \qquad \forall j \in \{1, \ldots, N\}$$

Consequently, we have $m_1 \leq z_j \leq M_1$ with $M_1 - m_1 = (1 - 2d)(M_0 - m_0)$.

Moreover, we easily see that $m_1 \geq m_0$ and $M_1 \leq M_0$. After t iterations, the coordinates of $\boldsymbol{P}^t\mathbf{y}$ *will be included between numbers* m_t *and* M_t, *satisfying:*

$$M_t - m_t = (1 - 2d)^t(M_0 - m_0)$$

and

$$m_0 \leq m_1 \leq \cdots \leq m_t \leq M_t \leq \cdots \leq M_1 \leq M_0.$$

The sequences $\{m_t\}_{t\geq 1}$ and $\{M_t\}_{t\geq 1}$ are therefore adjacent sequences and converge to the same limit u. Thus, we have:

$$\lim_{t\to\infty} \boldsymbol{P}^t\mathbf{y} = \begin{pmatrix} u \\ \vdots \\ u \end{pmatrix}$$

where u depends on $\mathbf{y}$. We can apply this equation to basis vectors $\mathbf{e}_1, \ldots, \mathbf{e}_N$. There exists, then, for every i, numbers π_i such that:

$$\lim_{t\to\infty} \boldsymbol{P}^t\boldsymbol{e}_i = \begin{pmatrix} \pi_i \\ \vdots \\ \pi_i \end{pmatrix}$$

Now $\boldsymbol{P}^t\boldsymbol{e}_i$ *is the i-th column of* $\boldsymbol{P}^t$, *so we have:*

$$\lim_{t\to\infty} \boldsymbol{P}^t = \begin{pmatrix} \pi_1 & \cdots & \pi_N \\ \vdots & \cdots & \pi_N \\ \pi_1 & \cdots & \pi_N \end{pmatrix}$$

By using equation [1.5] and given that the sum of the coordinates of $\boldsymbol{\pi}(0)$ is equal to 1, we have:

$$\lim_{t\to\infty} \boldsymbol{\pi}(t) = \lim_{t\to\infty} \boldsymbol{\pi}(0).\mathbf{P}^t = \boldsymbol{\pi}(0).\begin{pmatrix} \pi_1 & \cdots & \pi_N \\ \vdots & \cdots & \pi_N \\ \pi_1 & \cdots & \pi_N \end{pmatrix} = (\pi_1 \quad \cdots \quad \pi_N)$$

This equation is independent of $\boldsymbol{\pi}(0)$. We have demonstrated the first part.

Finally, it suffices to use the equation $\boldsymbol{\pi}(t+1) = \boldsymbol{\pi}(t).\mathbf{P}$. When the limit $\lim_{t\to\infty} \boldsymbol{\pi}(t)$ exists, we must have $\lim_{t\to\infty} \boldsymbol{\pi}(t) = \lim_{t\to\infty} \boldsymbol{\pi}(t+1) = \boldsymbol{\pi}$, so $\boldsymbol{\pi}$ must satisfy the equation $\boldsymbol{\pi} = \boldsymbol{\pi}\mathbf{P}$. ■

2.1.3. *Asymptotic behavior*

2.1.3.1. *Stationary distribution*

DEFINITION 2.13.– *A distribution (stochastic vector)* $\boldsymbol{\pi} = (\pi_0, \pi_1, \ldots, \pi_n)$ *is called stationary with respect to the stochastic matrix* **P** *if, and only if, it is constant over time, that is,*

$$\boldsymbol{\pi} = \boldsymbol{\pi}\mathbf{P} \qquad [2.10]$$

This condition is equally written $\boldsymbol{\pi}(\mathbf{P} - \mathbf{I}) = \mathbf{0}$. By noticing that 1 is a simple eigenvalue of **P**, we can say that $\boldsymbol{\pi}$ is the left eigenvector of **P** corresponding to the eigenvalue 1.

From Proposition 2.5 and Definition 2.13, we can deduce *the existence and uniqueness* of the stationary distribution of a regular Markov chain. However, in a general sense, a finite Markov chain allows for at least one stationary distribution. It is no longer necessarily true if the space of states is infinite.

If the Markov chain begins with the distribution $\boldsymbol{\pi}(0) = \boldsymbol{\pi}$, then the stochastic vector $\boldsymbol{\pi}(t)$ is constant for $t = 0, 1, 2, \ldots$ This is why it is called a *stationary distribution* (or an invariable distribution).

EXAMPLE.– Let there be a stochastic matrix:

$$\mathbf{P} = \begin{pmatrix} 1/2 & 0 & 1/2 \\ 1 & 0 & 0 \\ 1/4 & 1/4 & 1/2 \end{pmatrix}$$

The calculation is performed by solving the linear system $\boldsymbol{\pi} = \boldsymbol{\pi}\mathbf{P}$.

The linear equations of this system are not linearly independent because by adding them member by member, we obtain the trivial identity $1 = 1$. We must thus complete these equations with the condition $\pi_0 + \pi_1 + \cdots + \pi_n = 1$, which gives the following system of equations:

$$\begin{cases} \frac{1}{2}\pi_0 + \pi_1 + \frac{1}{4}\pi_2 = \pi_0 \\ \frac{1}{4}\pi_2 = \pi_1 \\ \frac{1}{2}\pi_0 + \frac{1}{2}\pi_2 = \pi_2 \\ \pi_0 + \pi_1 + \pi_2 = 1 \end{cases}$$

The calculations give $\boldsymbol{\pi} = (4/9, 1/9, 4/9)$.

2.1.3.2. *Ergodic theorem*

PROPOSITION 2.6 (Ergodic theorem).– *For an ergodic Markov chain, and for any initial distribution* $\boldsymbol{\pi}(0)$*, the average frequency of passing through any state* i *converges to* π_i *which is the* i*-th coordinate of the stationary distribution* $\boldsymbol{\pi}$ *of the Markov chain:*

$$\lim_{t\to\infty}\frac{1}{t}E\left(\sum_{n=0}^{t-1}1_{\{X_n=i\}}\right)=\pi_i \qquad \forall i\in\mathcal{X} \qquad [2.11]$$

In other words, the fraction of time spent in a state given by the process studied here is equal, over a long period of time, to the stationary probability of this state.

PROOF.–

Let $\boldsymbol{\Pi}$ be a matrix, all of whose coordinates are equal to $\boldsymbol{\pi}$ as used here in the proof of Proposition 2.5. Then, we have $\boldsymbol{\Pi}\mathbf{P}=\boldsymbol{\Pi}$, and we can also verify that $\mathbf{P}\boldsymbol{\Pi}=\boldsymbol{\Pi}$. It follows that:

$$(\mathbf{I}+\mathbf{P}+\cdots+\mathbf{P}^{t-1})(\mathbf{I}-\mathbf{P}+\boldsymbol{\Pi})=\mathbf{I}-\mathbf{P}^t+t\boldsymbol{\Pi}$$

Let us show that the matrix $\mathbf{I}-\mathbf{P}+\boldsymbol{\Pi}$ is invertible. Let $\mathbf{x}$ be a column vector such that $(\mathbf{I}-\mathbf{P}+\boldsymbol{\Pi})\mathbf{x}=\mathbf{0}$. Then, we have:

$$\mathbf{0}=\boldsymbol{\pi}(\mathbf{I}-\mathbf{P}+\boldsymbol{\Pi})\mathbf{x}=\boldsymbol{\pi}(\mathbf{I}-\mathbf{P})\mathbf{x}+\boldsymbol{\pi}\boldsymbol{\Pi}\mathbf{x}=\boldsymbol{\pi}\boldsymbol{\Pi}\mathbf{x}=\boldsymbol{\pi}\mathbf{x}$$

since $\boldsymbol{\pi}(\mathbf{I}-\mathbf{P})=0$ and $\boldsymbol{\pi}\boldsymbol{\Pi}=\boldsymbol{\pi}$ because of the fact that $\sum\pi_i=1$.

It follows that $\boldsymbol{\pi}\mathbf{x}=\mathbf{0}$, $\boldsymbol{\Pi}\mathbf{x}=\mathbf{0}$, and therefore $(\mathbf{I}-\mathbf{P})\mathbf{x}=\mathbf{0}$. Since $\mathbf{P}$ allows 1 as a simple eigenvalue, with the right eigenvector $\mathbf{1}$, this implies that $\mathbf{x}\parallel\mathbf{1}$, which is not possible unless $\mathbf{x}=\mathbf{0}$ since $\boldsymbol{\pi}\mathbf{x}=\mathbf{0}$ and all π_i are positive. The matrix $\mathbf{I}-\mathbf{P}+\boldsymbol{\Pi}$ is therefore indeed invertible.

Let $\mathbf{Z}=(\mathbf{I}-\mathbf{P}+\boldsymbol{\Pi})^{-1}$. Since $\boldsymbol{\pi}(\mathbf{I}-\mathbf{P}+\boldsymbol{\Pi})=\boldsymbol{\pi}$, we also have $\boldsymbol{\pi}=\boldsymbol{\pi}\mathbf{Z}$ and $\boldsymbol{\Pi}=\boldsymbol{\Pi}\mathbf{Z}$. After multiplication to the right by $\mathbf{Z}$, we have:

$$(\mathbf{I}+\mathbf{P}+\cdots+\mathbf{P}^{t-1})=(\mathbf{I}-\mathbf{P}^t+t\boldsymbol{\Pi})\mathbf{Z}=(\mathbf{I}-\mathbf{P}^t)\mathbf{Z}+t\boldsymbol{\Pi}\mathbf{Z}=(\mathbf{I}-\mathbf{P}^t)\mathbf{Z}+t\boldsymbol{\Pi}$$

Now, for any initial state i, we have:

$$\frac{1}{t}E\left(\sum_{n=0}^{t-1}1_{\{X_n=j\}}\right)=\frac{1}{t}\sum_{n=0}^{t-1}(\mathbf{P}^n)_{ij}=\left[\frac{1}{t}(\mathbf{I}-\mathbf{P}^t)\mathbf{Z}+\boldsymbol{\Pi}\right]_{ij}$$

Since the coordinates of matrix $\mathbf{P}^n$ are uniformly bounded by 1, this quantity converges to $[\mathbf{\Pi}]_{ij} = \pi_j$ when $t \to \infty$. In the same way, we obtain $\boldsymbol{\pi}(0)$ for any initial distribution. ■

2.1.4. *Holding time in a state*

PROPOSITION 2.7.– *Let p be the probability of leaving a certain state during a unit of time. For the Markovian process we are considering, the holding time T in this state is distributed according to the geometric average distribution* $1/p$*:*

$$\mathbb{P}(T = k) = (1 - p)^{k-1}p, \quad (k \geq 1) \qquad [2.12]$$

PROOF.–

For the proof, we consider the simplest case of a Markov chain with two states, represented in Figure 2.2.

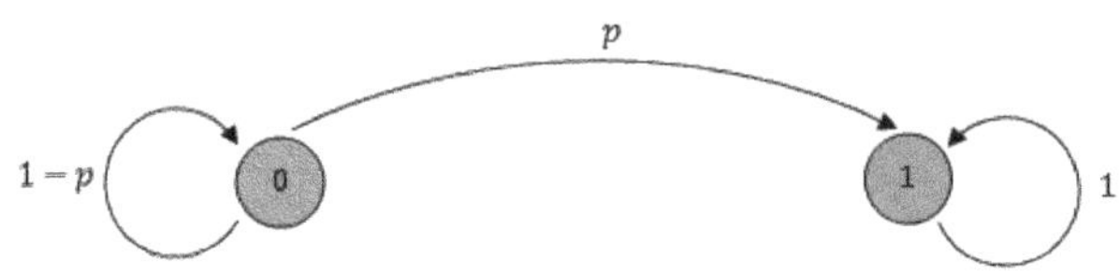

Figure 2.2. *Markov chain for a geometric distribution*

We denote by T the random variable equal to the holding time in state 0. We thus have:

$$\begin{aligned}
p_1 &:= \mathbb{P}(T = 1) = p \\
p_2 &:= \mathbb{P}(T = 2) = (1 - p)p \\
p_3 &:= \mathbb{P}(T = 3) = (1 - p)^2 p \\
&\dots \\
p_k &:= \mathbb{P}(T = k) = (1 - p)^{k-1}p, \quad k \geq 1
\end{aligned}$$

We arrive at the probability distribution for a random variable following a geometric distribution.

In a more general case, it is sufficient to group all the arrows leaving state i into just one with the probability $p = \sum_{j \neq i} P_{ij} = 1 - P_{ii}$ as in Figure 2.2, and in the same way, we arrive at this geometric distribution with parameter p. ■

The average holding time T in a state is thus equal to:

$$E(T) = \frac{1}{p} \quad [2.13]$$

This formula is very intuitive. A person who has one chance out of p to succeed in an exam will have to, on average, take it p times.

2.1.5. *Time-reversible chain*

Up until now, we have only considered Markov chains $X(t) \in \mathcal{X}$ indexed by time $t \in \mathcal{T} = \mathbb{N}$. We can construct a Markov chain $X(t) \in \mathcal{X}$ indexed by time $t \in \mathcal{T} = \mathbb{Z}$ from a stable chain and in a steady state. In this case, the evolution of the chain begins at time $t = -\infty$, as shown in Figure 2.3.

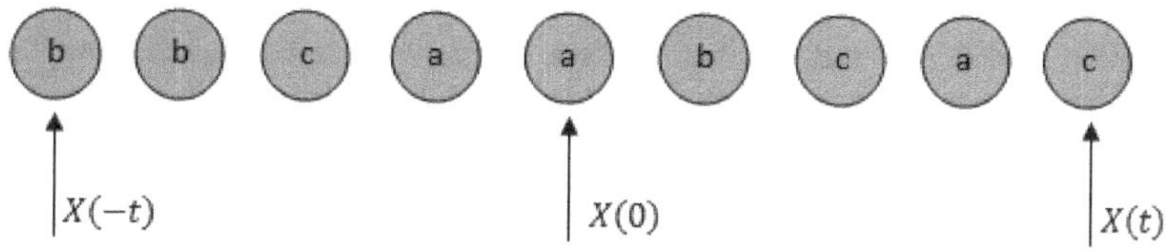

Figure 2.3. *Time-reversible Markov chain*

DEFINITION 2.14.– *We define the time-reversible chain from chain $X(t)$ by:*

$$\tilde{X}(t) = X(-t), \qquad t \in \mathbb{N} \quad [2.14]$$

Let $\widetilde{\mathbf{P}} = (\tilde{p}_{ij}) = (\tilde{p}(i,j))$ be the transition matrix for this time-reversible chain.

PROPOSITION 2.8.– *The transition matrix $\widetilde{\mathbf{P}}$ of the time-reversible chain is equal to:*

$$\tilde{P}_{ij} = \frac{\pi_j}{\pi_i} P_{ji}, \quad \forall i,j \in \mathcal{X} \quad [2.15]$$

PROOF.–

We can calculate the coordinate $\tilde{P}_{ij}$ from the transitions between states i and j:

$$\begin{aligned}\tilde{P}_{ij} &= \mathbb{P}(\tilde{X}(1) = j \mid \tilde{X}(0) = i) \\ &= \mathbb{P}(X(0) = j \mid X(1) = i)\end{aligned}$$

$$= \frac{\mathbb{P}(X(0) = j)}{\mathbb{P}(X(1) = i)} \mathbb{P}(X(1) = i \mid X(0) = j)$$
$$= \frac{\pi_j}{\pi_i} P_{ji}, \quad \forall i, j \in \mathcal{X} \qquad \blacksquare$$

2.1.6. *Reversible Markov chains*

DEFINITION 2.15.– *A stable Markov chain $X(t)$ is called reversible if it follows the same distribution as its time-reversible chain $\tilde{X}(t)$, that is, if:*

$$\forall i, j \in \mathcal{X}, \qquad \tilde{P}_{ij} = P_{ij} \qquad [2.16]$$

From equation [2.15], we can deduce that for a reversible Markov chain:

$$\forall i, j \in \mathcal{X}, \qquad \pi_i P_{ij} = \pi_j P_{ji} \qquad [2.17]$$

Equation [2.17] presents what are called *local balance equations*. They mean that, in a steady state, the frequency of transitions from state i to state j must be equal to that of state j to state i, whatever the states $i, j \in \mathcal{X}$ may be.

More generally, we can say a Markov chain is reversible if a non-zero distribution $\boldsymbol{\pi}$ exists which satisfies the local balance equations (we do not impose stability).

2.1.7. *Kolmogorov's criterion*

PROPOSITION 2.9 (Kolmogorov's criterion).– *A Markov chain is reversible if, and only if, its transition graph is symmetrical and, for any cycle in this graph, that is, any path $i_0, i_1, \ldots, i_l$ in the graph such that $i_l = i_0$, where l is the length of the cycle, the product of the transition probabilities is the same in both directions of the cycle:*

$$p(i_0, i_1)p(i_1, i_2) \ldots p(i_{l-1}, i_l) = p(i_l, i_{l-1}) \ldots p(i_2, i_1)p(i_1, i_0) \qquad [2.18]$$

where $p(i, j)$ designates P_{ij}.

PROOF.–

A solution for local balance equations is thus given by $\pi_{i_0} = \alpha$, where $i_0 \in \mathcal{X}$ designates an arbitrary state and α a strictly positive constant, and:

$$\forall i \in \mathcal{X}, i \neq i_0, \pi_{i_0} = \alpha \frac{p(i_0, i_1)p(i_1, i_2) \dots p(i_{l-1}, i)}{p(i_1, i_0)p(i_2, i_1) \dots p(i, i_{l-1})}$$

where $i_0, i_1, \dots, i_{l-1}, i$ designates any path from i_0 to i in the transition graph. This expression does not depend on the path chosen and for every pair of states $i, j \in \mathcal{X}$ connected by an arc in the transition graph, we have:

$$\pi_i \frac{p(i,j)}{p(j,i)} = \frac{p(i_0, i_1)p(i_1, i_2) \dots p(i_{l-1}, i)\, p(i,j)}{p(i_1, i_0)p(i_2, i_1) \dots p(i, i_{l-1})\, p(j,i)} = \pi_j$$

which corresponds to the local balance equations.

Reciprocally, the local balance equations imply the symmetry of the transition graph since:

$$\frac{p(i_0, i_1)}{p(i_1, x_0)} \frac{p(i_1, i_2)}{p(i_2, x_1)} \cdots \frac{p(i_{l-1}, i)}{p(i, i_{l-1})} = \frac{\pi_{i_1}}{\pi_{i_0}} \frac{\pi_{i_2}}{\pi_{i_1}} \cdots \frac{\pi_{i_l}}{\pi_{i_{l-1}}} = 1$$

This equation justifies the utility of reversible Markov chains, for which the solution for balance equations is explicit. ■

2.2. Markov chains in continuous time

2.2.1. *Definitions*

In section 2.2, we consider $\mathcal{X} = \{0, \dots, n\} \subset \mathbb{N}$ *and* $\mathcal{T} = \mathbb{R}^+$, *that is, the possible state of this system is an integer between* 0 and n and that time t is a positive real number or zero.

We always denote by $\pi_i(t)$ the probability that the system will be in state $X(t) = i$ at instant t, $\pi_i(t) = \mathbb{P}(X(t) = i)$. $\boldsymbol{\pi}(t) = \big(\pi_0(t), \pi_1(t), \dots, \pi_n(t)\big)$ is the stochastic vector at any instant t with coordinates $\pi_i(t)$, $t \in \mathbb{R}^+$.

2.2.1.1. *Markov chains in continuous time*

The analysis of the process begins at time 0, and time t passes in a continuous manner. $X(t) \in \mathcal{X} = \{0, \dots, n\}$ *designates the state of the system at time* t.

The instants when states change $t_1, t_2, \ldots$ *are random instants in time:*

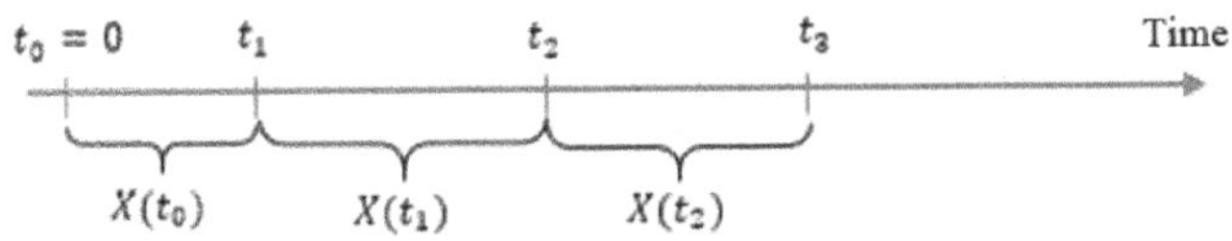

Figure 2.4. *Instants of changes of state*

Let there be three consecutive instants in which there have been changes of state:

– r: past instant ($r \geq 0$), with $X(r) = l$;

– s: present instant, or current instant ($s > r$), with $X(s) = i$;

– $s + t$: future instant ($t > 0$), with $X(s + t) = j$.

DEFINITION 2.16 (Weak Markov property).– *A random process in continuous time* $\{X(t)\}_{t \in \mathbb{R}^+}$ *with values in a discrete set* $\mathcal{X}$ *is a Markov chain in continuous time if it is memoryless in the sense that for any time* $r \geq 0,\ \ s > r,\ t > 0$*, for any state* $i, j, l \in \mathcal{X}$*:*

$$\begin{aligned}\mathbb{P}(X(s+t) = j | X(s) = i, X(r) = l) \\ = \mathbb{P}(X(s+t) = j | X(s) = i)\end{aligned} \qquad [2.19]$$

As with this discrete formalism, the probabilities $\mathbb{P}(X(s+t) = j | X(s) = i)$ are transition probabilities.

DEFINITION 2.17.– *The Markov chain is called homogeneous if the definite probabilities in equation* [2.19] *are independent of time* s:

$$\mathbb{P}(X(s+t) = j | X(s) = i) = \mathbb{P}(X(t) = j | X(0) = i) \quad \forall s > 0 \qquad [2.20]$$

$P_{ij}(t) = \mathbb{P}(X(t) = j | X(0) = i)$ is the transition probability function in continuous time.

2.2.1.2. *Holding time in a state*

PROPOSITION 2.10.– *The holding time in a given state for a homogeneous Markov chain in continuous time follows an exponential distribution.*

PROOF.–

We will denote by T_i the time spent by the system in state i before changing to another state j.

Let us suppose that the process entered state i at instant s. For any time $t > 0$, the event $T_i > t$ is equivalent to $X(t') = i$ for any $t' \in (s, s+t)$. Thus:

$$\mathbb{P}(T_i > t) = \mathbb{P}(T_i > s + t | T_i > s)$$

This is a memoryless property of time T_i. The only random variable satisfying this property is the random variable following an exponential distribution according to Proposition 1.1. ■

By denoting with q_i the parameter of the exponential distribution followed by the holding time T_i, we have:

$$\mathbb{P}(T_i \leq t) = 1 - e^{-q_i t} \qquad [2.21]$$

Its average is equal to:

$$E(T_i) = \frac{1}{q_i} \qquad [2.22]$$

REMARK.– A Markov chain in continuous time can thus be described as follows:

– the random variable T_i of the holding time in state i follows an exponential distribution with parameter q_i;

– when the process leaves state i, it passes to state j with a probability p_{ij};

– the next state visited after i is independent of the time spent in state i.

2.2.1.3. *Chapman–Kolmogorov equation*

The transition probabilities satisfy a property similar to that of Markov chains in discrete time. Let there be times $s,\ t \in \mathbb{R}^+$ such that $s < t$.

We denote $P_{ij}(s,t) = \mathbb{P}(X(t) = j \,| X(s) = i)$.

PROPOSITION 2.11 (Chapman–Kolmogorov equation).– *The transition probabilities between any two states i and j satisfy the following equation:*

$$P_{ij}(s,t) = \sum_{k=0}^{n} P_{ik}(s,r) P_{kj}(r,t) \qquad [2.23]$$

PROOF.–

The demonstration is identical to that of Proposition 2.2. ■

2.2.2. Evolution over time

2.2.2.1. *Transition intensities*

Recall that $P_{ij}(t) = \mathbb{P}(X(t) = j|X(0) = i)$ is the transition probability function in continuous time.

PROPOSITION 2.12.– *The parameter q_i of the holding time in state i satisfies:*

$$q_i = -\left.\frac{dP_{ii}}{dt}\right|_{t=0} = \lim_{t\to 0}\frac{1 - P_{ii}(t)}{t} \qquad [2.24]$$

PROOF.–

We have $P_{ii}(t) = \mathbb{P}(X(t) = i|X(0) = i)$: this is the probability of finding the system in state i again at instant t. Consequently, it is equal to $\mathbb{P}(T_i > t)$.

$$\mathbb{P}(T_i > t) = e^{-q_i t} = P_{ii}(t)$$

Its derivative at instant $t = 0$ gives us:

$$-\left.\frac{dP_{ii}}{dt}\right|_{t=0} = q_i e^{-q_i t}|_{t=0} = q_i \quad ■$$

q_i is called the *transition intensity* from state i.

We thus define the transition intensity in the same way from i to j such that:

$$q_{ij} = \left.\frac{dP_{ij}}{dt}\right|_{t=0} = \lim_{t\to 0}\frac{P_{ij}(t)}{t} \qquad [2.25]$$

We can interpret it as the average number of times that the process passes from state i to state j per unit of time.

PROPOSITION 2.13.– *Transition intensities satisfy:*

$$q_i = \sum_{\substack{j=0 \\ j\neq i}}^{n} q_{ij} \qquad [2.26]$$

PROOF.–

Proposition 2.13 easily derives from the definition of transition probabilities $\sum_{j=0}^{n} p_{ij}(t) = 1$ and transition intensities. ■

The *transition graph* of a Markov chain in continuous time is semi-identical to that of a Markov chain in discrete time, as shown in Figure 2.5. The only difference is in the weight of the arcs connecting the states: we use transition intensities from state i to state j, and we omit the transition intensities of state i towards itself given that we can use equation [2.26] to determine them.

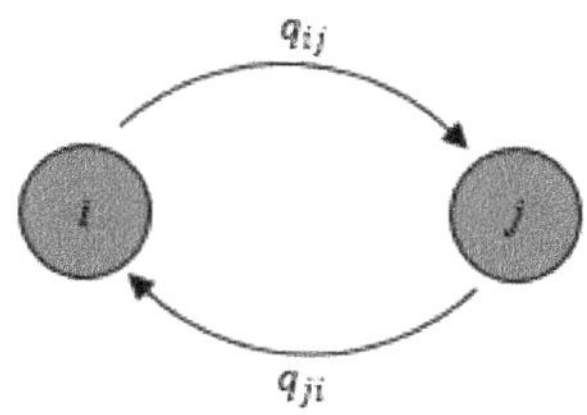

Figure 2.5. *Transition graph of a Markov chain in continuous time*

If we denote by T_{ij} the time spent in state i before a transition to state j, we can find, as we did with Proposition 2.12, that T_{ij} follows an exponential distribution with parameter q_{ij}. When the transition happens, the probability that it will happen in the direction of state j is:

$$P_{ij} = \frac{q_{ij}}{q_i} \qquad [2.27]$$

2.2.2.2. *Infinitesimal stochastic generator*

DEFINITION 2.18.– *A real square matrix* **A** *is called an infinitesimal stochastic generator if:*

– *the sum of the coordinates in any one row is zero;*

– *the coordinates on the diagonal are negative or zero and the others are positive or zero.*

EXAMPLE.– Matrix **A** is an infinitesimal stochastic generator:

$$\mathbf{A} = \begin{pmatrix} -2 & 2 & 0 \\ 1 & -3 & 2 \\ 0 & 1 & -1 \end{pmatrix}$$

PROPOSITION 2.14.– *Matrix* $\boldsymbol{A} = (a_{ij})$, *formed by transition intensities such that* $a_{ij} = q_{ij}$ $(\forall i \neq j)$ *and* $a_{ii} = -q_i$, *is an infinitesimal stochastic generator.*

PROOF.–

From equation [2.26], we find that the sum of coordinates of each row is equal to 0 and, as per its very definition, we find that the diagonal coordinates a_{ii} are negative or zero, and the others a_{ij} are positive or zero. ■

Proposition 2.14 allows us to define a Markov chain in continuous time from its matrix **A** of transition intensities, called an infinitesimal stochastic generator of the chain.

2.2.2.3. *State equation*

PROPOSITION 2.15.– *The stochastic vector* $\boldsymbol{\pi}(t)$ *of a Markov chain in continuous time of infinitesimal stochastic generator* **A** *satisfies, at any instant* t*, the ordinary differential equation of the form:*

$$\frac{d\boldsymbol{\pi}(t)}{dt} = \boldsymbol{\pi}(t)\mathbf{A} \tag{2.28}$$

PROOF.–

We consider a random dynamic system (process) whose evolution is governed by a Markov chain in discrete time, the time step between two "ticks" of a clock supposedly being infinitely small (positive). Such an infinitely small positive real number will be denoted by ε. The transition matrix denoted by $\mathbf{P}(\varepsilon)$ is nearly identical because we suppose that the system "evolves very little" during this time ε.

From limits [2.24] and [2.25], we can write that:

$$\mathbf{P}(\varepsilon) = \mathbf{I} + \varepsilon\mathbf{A} + \mathbf{o}(\varepsilon), \qquad \varepsilon \to 0^+$$

We thus have:

$$\begin{aligned}\boldsymbol{\pi}(t+\varepsilon) &= \boldsymbol{\pi}(t)\mathbf{P}(\varepsilon)\\ &= \boldsymbol{\pi}(t)(\mathbf{I} + \varepsilon\mathbf{A})\\ &= \boldsymbol{\pi}(t) + \varepsilon\boldsymbol{\pi}(t)\mathbf{A}\\ \frac{\boldsymbol{\pi}(t+\varepsilon) - \boldsymbol{\pi}(t)}{\varepsilon} &= \boldsymbol{\pi}(t)\mathbf{A}\end{aligned}$$

When $\varepsilon \to 0^+$, the growth rate on the left side of the last equation tends towards the derivative $\frac{d\pi(t)}{dt} = \dot{\boldsymbol{\pi}}(t)$.■

The ordinary differential equation [2.28] is called the *state equation* of the Markov chain in continuous time.

2.2.3. *Resolving the state equation*

2.2.3.1. *Solution to the state equation*

PROPOSITION 2.16.– The state equation $\dot{\boldsymbol{\pi}}(t) = \boldsymbol{\pi}(t)\mathbf{A}$ allows as a solution the vector $\boldsymbol{\pi}(t) = \boldsymbol{\pi}(0)\exp(t\mathbf{A})$ with:

$$\exp(t\mathbf{A}) = 1 + t\mathbf{A} + \frac{1}{2!}t^2\mathbf{A}^2 + \frac{1}{3!}t^3\mathbf{A}^3 + \cdots \qquad [2.29]$$

$\mathbf{A}$ is an infinitesimal stochastic generator, so the matrix $\exp(t\mathbf{A})$ is, at any instant t, a stochastic matrix.

PROOF.–

By deriving the Taylor[4] development term by term, we obtain $\frac{d}{dt}\big(\exp(t\mathbf{A})\big) = \exp(t\mathbf{A}).\mathbf{A}$. From this we can deduce that:

$$\dot{\boldsymbol{\pi}}(t) = \boldsymbol{\pi}(0)\frac{d}{dt}\big(\exp(t\mathbf{A})\big) = \boldsymbol{\pi}(0)\exp(t\mathbf{A}).\mathbf{A} = \boldsymbol{\pi}(t)\mathbf{A}$$

Thus, the row vector $\boldsymbol{\pi}(t)$ satisfies the state equation; it is the solution for equation [2.28].

Let us show that the matrix $\mathbf{P}(t) := \exp(t\mathbf{A})$ is stochastic. Matrix $\mathbf{P}(t)$ satisfies differential equation $\dot{\mathbf{P}}(t) = \mathbf{P}(t)\mathbf{A}$ with the initial condition $\mathbf{P}(0) = \mathbf{I}$. Let $\mathbf{u}$ be the column vector whose $n+1$ coordinates are equal to 1. Each line of the matrix $\mathbf{A}$ has a sum of zero; therefore, $\mathbf{Au} = \mathbf{0}$. We can deduce from this that $\dot{\mathbf{P}}(t)\mathbf{u} = \mathbf{P}(t)\mathbf{Au} = \mathbf{0}$. The vector $\mathbf{P}(t)\mathbf{u}$ whose derivative is zero is thus constant. This constant evaluated at instant $t = 0$ has the value of $\mathbf{P}(0)\mathbf{u} = \mathbf{Iu} = \mathbf{u}$. Therefore, $\mathbf{P}(t)\mathbf{u} = \mathbf{u}$ whatever t may be and it follows that the sum of coordinates in any one row of $\mathbf{P}(t)$ is equal to 1. ■

2.2.3.2. *Calculating the exponential of a square matrix*

We can also use the eigendecomposition of matrix $\mathbf{A}$ into $\mathbf{A} = \mathbf{QDQ}^{-1}$, where $\mathbf{Q}$ is the matrix whose i-th column is the eigenvector of $\mathbf{A}$ and $\mathbf{D}$ is the diagonal matrix whose diagonal coefficients are eigenvalues λ_i: $\mathbf{D} = \text{diag}(\lambda_0, \lambda_1, \ldots, \lambda_n)$. In this case, we have $e^{t\mathbf{A}} = \mathbf{Q}e^{t\mathbf{D}}\mathbf{Q}^{-1}$ with $e^{t\mathbf{D}} = \text{diag}(e^{\lambda_0 t}, e^{\lambda_1 t}, \ldots, e^{\lambda_n t})$.

From this, we deduce that each coordinate of matrix $\exp(t\mathbf{A})$ is a linear combination of exponential functions $\exp(\lambda_i t)$, that is,

4 Brook Taylor (1865–1731), English mathematician.

$$(e^{t\mathbf{A}})_{i,j} = \sum_{k=0}^{n} a_{i,j,k} e^{\lambda_k t} \qquad [2.30]$$

where $a_{i,j,k}$ are complex constants.

In the case where several eigenvalues of **A** are equal (the characteristic polynomial of **A** allows for at least one double root), it is possible that matrix **A** is not diagonalizable.

2.2.3.3. *Discretization of a Markov chain in continuous time*

The stochastic matrix $\mathbf{P}(\varepsilon) := \exp(\varepsilon\mathbf{A})$ calculated for an infinitely small time ε allows for the limited expansion $\mathbf{P}(\varepsilon) = \mathbf{I} + \varepsilon\mathbf{A} + \mathbf{o}(\varepsilon)$ at around 0.

With the formula $\boldsymbol{\pi}(t) = \boldsymbol{\pi}(0)\exp(t\mathbf{A})$ from Proposition 2.16, we arrive at:

$$\boldsymbol{\pi}(t+1) = \boldsymbol{\pi}(t)\exp(\mathbf{A}) \qquad [2.31]$$

We can deduce that the Markov chain in discrete time whose transition matrix is equal to $\mathbf{P} = \exp(\mathbf{A})$ has the same behavior, for integer values of time, as the Markov chain in continuous time whose infinitesimal stochastic generator is matrix **A**.

2.2.4. *Asymptotic behavior*

2.2.4.1. *Stationary distribution*

DEFINITION 2.19.– *A distribution (stochastic vector)* $\boldsymbol{\pi} = (\pi_0, \pi_1, \ldots, \pi_n)$ *is called stationary with respect to the infinitesimal stochastic generator* **A** *of transition intensities if, and only if, it is constant over time, that is,*

$$\boldsymbol{\pi}\mathbf{A} = \mathbf{0} \qquad [2.32]$$

This definition comes from equation [2.28]: $\dot{\boldsymbol{\pi}}(t) = \boldsymbol{\pi}(t)\mathbf{A}$. The stochastic vector $\boldsymbol{\pi}(t)$ is invariable over time if $\dot{\boldsymbol{\pi}}(t) = \mathbf{0}$, that is, $\boldsymbol{\pi}(t)\mathbf{A} = \mathbf{0}$.

2.2.4.2. *Ergodic theorem*

We can obtain the equivalent of the ergodic theorem for the case of continuous time.

PROPOSITION 2.17 (Ergodic theorem).– *For an ergodic Markov chain in continuous time, and for any initial distribution* $\boldsymbol{\pi}(0)$, *the average frequency of passing through* *any state i converges to π_i, which is the i-th coordinate of the stationary distribution* $\boldsymbol{\pi}$ *of the Markov chain:*

$$\lim_{t\to\infty}\frac{1}{t}\int_0^t 1_{\{X(s)=i\}}\,ds = \pi_i \qquad \forall i \in \mathcal{X} \tag{2.33}$$

In other words, the fraction of time spent in a state given by the process studied here is equal, over a long period of time, to the stationary probability of this state.

PROOF.– Accepted.

2.3. Birth and death process

2.3.1. *Definition*

Let there be a queue containing between 0 and n people. The number of people present in the queue at instant t is the state of the process. Take any positive numbers $(\lambda_0, \lambda_1, \ldots, \lambda_{n-1})$ and $(\mu_1, \mu_2, \ldots, \mu_n)$. We can suppose that during an infinitely small interval of time ε, given that there are k people present in the queue:

– the probability that one person will arrive in the queue is $\lambda_k\varepsilon + o(\varepsilon)$;

– the probability that one person will leave the queue is $\mu_k\varepsilon + o(\varepsilon)$;

– other events have a probability of $o(\varepsilon)$.

These hypotheses amount to positing that the transition matrix $\mathbf{P}(\varepsilon)$ calculated for the interval of time ε has the form:

$$\mathbf{P}(\varepsilon) = \begin{pmatrix} 1-\lambda_0\varepsilon & \lambda_0\varepsilon & 0 & 0 & 0 \\ \mu_1\varepsilon & 1-(\lambda_1+\mu_1)\varepsilon & \lambda_1\varepsilon & 0 & 0 \\ 0 & \mu_2\varepsilon & 1-(\lambda_2+\mu_2)\varepsilon & \lambda_2\varepsilon & 0 \\ \vdots & \ddots & \ddots & \ddots & \ddots \\ 0 & 0 & 0 & \mu_n\varepsilon & 1-\mu_n\varepsilon \end{pmatrix} \tag{2.34}$$

2.3.2. *Infinitesimal stochastic generator*

The infinitesimal stochastic generator $\mathbf{A}$ satisfies, by definition, $\mathbf{P}(\varepsilon) = \mathbf{I} + \varepsilon\mathbf{A} + \mathbf{o}(\varepsilon)$:

$$\mathbf{A} = \begin{pmatrix} -\lambda_0 & \lambda_0 & 0 & 0 & 0 \\ \mu_1 & -(\lambda_1+\mu_1) & \lambda_1 & 0 & 0 \\ 0 & \mu_2 & -(\lambda_2+\mu_2) & \lambda_2 & 0 \\ \vdots & \ddots & \ddots & \ddots & \ddots \\ 0 & 0 & 0 & \mu_n & -\mu_n \end{pmatrix} \quad [2.35]$$

The process is represented by the diagram in Figure 2.6.

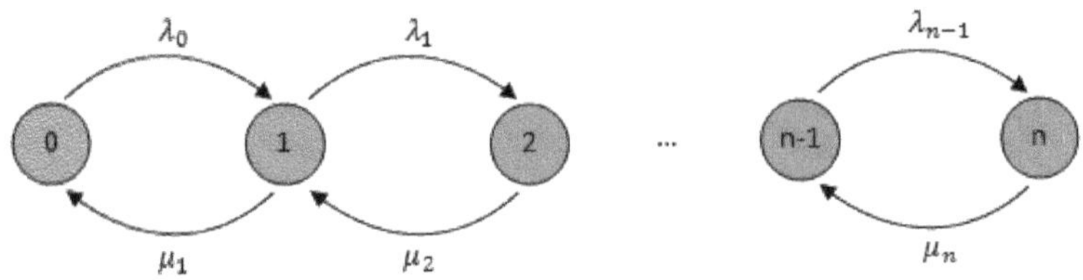

Figure 2.6. *Birth and death process*

2.3.3. *Stationary distribution*

PROPOSITION 2.18.– *The stationary distribution* $\boldsymbol{\pi}$ *of this birth and death process has the following coordinate:*

$$\pi_k = \frac{\frac{\lambda_0\lambda_1 \dots \lambda_{k-1}}{\mu_1\mu_2 \dots \mu_k}}{1 + \frac{\lambda_0}{\mu_1} + \frac{\lambda_0\lambda_1}{\mu_1\mu_2} + \dots + \frac{\lambda_0\lambda_1 \dots \lambda_{n-1}}{\mu_1\mu_2 \dots \mu_n}} \quad [2.36]$$

PROOF.–

The stationary distribution $\boldsymbol{\pi}$ is calculated from the solution to equation [2.32], $\boldsymbol{\pi}\mathbf{A} = \mathbf{0}$.

We obtain the following system of equations:

$$\begin{cases} -\lambda_0\pi_0 + \mu_1\pi_1 = 0 \\ \lambda_0\pi_0 - (\lambda_1+\mu_1)\pi_1 + \mu_2\pi_2 = 0 \\ \lambda_1\pi_1 - (\lambda_2+\mu_2)\pi_2 + \mu_3\pi_3 = 0 \\ \dots \\ \lambda_{n-2}\pi_{n-2} - (\lambda_{n-1}+\mu_{n-1})\pi_{n-1} + \mu_n\pi_n = 0 \\ \lambda_{n-1}\pi_{n-1} - \mu_n\pi_n = 0 \end{cases}$$

These equations can be reduced. It is more economical to write the so-called *balance* equations directly:

$$\lambda_0\pi_0 = \mu_1\pi_1,\ \ \lambda_1\pi_1 = \mu_2\pi_2,\ \ \lambda_2\pi_2 = \mu_3\pi_3, \dots, \lambda_{n-1}\pi_{n-1} = \mu_n\pi_n$$

as well as:

$$\pi_1 = \frac{\lambda_0}{\mu_1}\pi_0,\ \ \pi_2 = \frac{\lambda_1}{\mu_2}\pi_1,\ \ \pi_3 = \frac{\lambda_2}{\mu_3}\pi_2, \dots,\ \ \pi_n = \frac{\lambda_{n-1}}{\mu_n}\pi_{n-1}$$

and then:

$$\pi_k = \frac{\lambda_0\lambda_1 \dots \lambda_{k-1}}{\mu_1\mu_2 \dots \mu_k}\pi_0\,, \qquad k = 1,2,\dots,n$$

Considering the equation $\sum_{k=0}^{n}\pi_k = 1$, the value of π_0 is given by the formula:

$$\frac{1}{\pi_0} = 1 + \frac{\lambda_0}{\mu_1} + \frac{\lambda_0\lambda_1}{\mu_1\mu_2} + \dots + \frac{\lambda_0\lambda_1 \dots \lambda_{n-1}}{\mu_1\mu_2 \dots \mu_n}$$

Finally, we obtain the stationary distribution $\boldsymbol{\pi} = (\pi_0, \pi_1, \dots, \pi_n)$:

$$\pi_k = \frac{\dfrac{\lambda_0\lambda_1 \dots \lambda_{k-1}}{\mu_1\mu_2 \dots \mu_k}}{1 + \dfrac{\lambda_0}{\mu_1} + \dfrac{\lambda_0\lambda_1}{\mu_1\mu_2} + \dots + \dfrac{\lambda_0\lambda_1 \dots \lambda_{n-1}}{\mu_1\mu_2 \dots \mu_n}} \qquad ■$$

2.4. Exercises

EXERCISE 2.1.– *Ergodic*

Let there be a matrix $\mathbf{P} = \begin{pmatrix} 1/2 & 1/2 & 0 \\ 1/3 & 1/3 & 1/3 \\ 0 & 1/2 & 1/2 \end{pmatrix}$.

1) Verify that **P** is a transition matrix for a Markov chain in discrete time.

2) Draw the transition graph for this chain.

3) Calculate $\mathbf{P}^2$.

4) What can we deduce about the ergodicity of the corresponding Markov chain?

EXERCISE 2.2.– *Upper triangular matrix*

Let there be a matrix $\mathbf{P} = \begin{pmatrix} p^2 & 2p(1-p) & (1-p)^2 \\ 0 & p & 1-p \\ 0 & 0 & 1 \end{pmatrix}$.

1) Verify that $\mathbf{P}$ is a transition matrix for a Markov chain in discrete time.

2) Noting that all the powers of $\mathbf{P}$ are upper triangular, show that matrix $\mathbf{P}$ does not satisfy the conditions of Proposition 2.4.

3) Show that the distribution limit exists nonetheless.

4) Calculate the stationary distribution of the corresponding Markov chain.

5) What can we deduce about the ergodicity of this chain?

EXERCISE 2.3.– *Stationary distribution*

Calculate the stationary distribution of the Markov chains whose matrix $\mathbf{P}$ is equal to:

1) $\mathbf{P} = \begin{pmatrix} 1/2 & 0 & 1/2 \\ 1 & 0 & 0 \\ 1/4 & 1/4 & 1/2 \end{pmatrix}$

2) $\mathbf{P} = \begin{pmatrix} p(2-p) & (1-p)^2 \\ p & 1-p \end{pmatrix}, \quad 0 \leq p \leq 1$

EXERCISE 2.4.– *Non-existence of a stationary distribution*

Let there be a matrix $\mathbf{P} = \begin{pmatrix} 0 & 1 \\ 1 & 0 \end{pmatrix}$.

Show that for $\boldsymbol{\pi}(0) \neq \left(\frac{1}{2}, \frac{1}{2}\right)$, the limit $\lim_{t\to\infty} \boldsymbol{\pi}(t)$ does not exist.

EXERCISE 2.5.– *Bee*

A bee gathers pollen from three fields of flowers numbered 1, 2 and 3. It visits these three fields in a cyclical manner. However, after field 2, it returns to field 1 with a probability p (it thus continues its cycle to field 3 with a probability of $1-p$). Given that it stays the same amount of time in each field before going to the next:

1) Show that the field visited by the bee can be modeled by a Markov chain. Calculate its transition matrix and draw its transition graph.

2) Calculate the fraction of time it spends in each field.

EXERCISE 2.6.– *Traffic info*

On a road, three trucks out of four are followed by a car and one car out of five is followed by a truck. What is the proportion of cars on the road?

EXERCISE 2.7.– *Umbrella*

Rakoto only owns three umbrellas. Every day, he goes to work in the morning and comes back home in the evening. For each journey, he takes an umbrella with him if it is raining, and if there is at least one located where he finds himself (at his home or at his office). He does not take an umbrella with him if it is not raining. We can suppose that the probability that it will rain at the start of each journey is $1/3$, and that it is independent of the forecast during all the other journeys. Let X_n be the number of umbrellas that Rakoto owns in a given place before starting the n-th journey.

1) Show that X_n is a Markov chain. Provide its transition matrix and draw its transition graph .

2) What is the probability that, after a large number of journeys, Rakoto will not have an umbrella available in his location when leaving?

3) What is the probability that he will unwittingly get wet, that is, he will not have an available umbrella when it is raining at his time of departure?

EXERCISE 2.8.– *Traffic*

A road has four lanes. Each one can be crossed in one second. The probability that a car will arrive during a particular second is 0.8:

1) Draw the graph of the system.

2) Provide the transition matrix.

3) Calculate the average time to cross the four lanes.

EXERCISE 2.9.– *Printer*

Consider how a printer works. It can be in three distinct states:

– state 1: waiting to print a character;

– state 2: printing a character;

– state 3: interruption after receiving a control character.

When the printer is waiting, it receives a character to print with a probability of 0.80.

When it is printing, it receives:

– a normal character with a probability of 0.95 (common character from the file to print);

– an end of file character with a probability of 0.04 (the printer returns to the waiting state);

– an interruption character with a probability of 0.01 (the printer then goes into state 3).

When the printer is in state 3, it returns to the waiting state with a probability of 0.3, otherwise it remains in state 3.

1) Show that this system can be modeled by a Markov chain with three states.

2) Draw the graph associated with this chain and provide its transition matrix.

3) What is the average time of an interruption?

4) Write the equilibrium equations and show that this chain is ergodic. Calculate the associated stationary probabilities.

5) In a steady state, what is the rate of use of the printer?

EXERCISE 2.10.– *Fleet of buses*

Consider a fleet of N buses. Each vehicle breaks down independently of the others with a rate of μ and is sent to the garage. The mechanics in the garage can only repair one vehicle at a time, and the duration of their work is distributed according to an exponential distribution with λ.

1) Draw the transition graph for the Markov chain that models this system.

2) Calculate the stationary distribution of the number of buses in service.

3) What is the average number of buses in service?

EXERCISE 2.11.– *Travel*

A businessman travels between Antananarivo, Mahajanga and Antsiranana (large cities in Madagascar). The amount of time he spends in each city follows an exponential distribution, on average $1/4$ of the months in Antananarivo and Mahajanga and $1/5$ of the months in Antsiranana. If he is in Antananarivo, he will go to Mahajanga or Antsiranana with a probability of $1/2$. If he is in Mahajanga, he will go to Antananarivo with a probability of $3/4$ and to Antsiranana with a probability of $1/4$. After visiting Antsiranana, he always returns to Antananarivo.

1) Provide the generator for the Markov chain describing the businessman's itinerary.

2) Determine the fractions of time he travels to each city.

3) How many trips does he take on average from Antananarivo to Mahajanga per year?

EXERCISE 2.12.– *Stock*

A small computer store can have at most three computers in stock. Customers arrive with a rate of two customers per week. If at least one computer is in stock, the customer buys it. If there is at most one computer, the manager of the store orders two new computers, which are delivered after a certain amount of time following an exponential distribution, on average one week.

1) Provide the generator of the Markov chain describing the number of computers in stock.

2) Determine the stationary distribution.

3) What is the rate of sale of computers?

EXERCISE 2.13.– *Call center*

Calls arrive in a call center following a Poisson process with rate λ. There are s available operators to respond to calls. We suppose that the durations of these calls are independent and identically distributed with an exponential distribution whose average is $1/\mu$. A call that arrives when all operators are busy is refused.

1) Determine the stationary distribution of the number of busy operators.

2) Calculate the probability that a call will be rejected (the result is known as the first Erlang formula).

PART 2

Queues

3

Common Queues

Waiting doesn't wait for anything.

Maurice Blanchot (1907–2003)

Queues are everywhere in our everyday life: in banks, at the post office, in traffic, at stock depots, on the telephone, at gas stations… They occupy an important place in human life ensuring that resources can be well-managed.

This chapter is dedicated to the study of queue theory. We elucidate mathematical models for these queues that can help improve their use in everyday life, and especially in the field of telecommunication where resources are very scarce.

3.1. Arrival process of customers in a queue

3.1.1. *The Poisson process*

As was explained in the first chapter, the Poisson process is important for describing the occurrence of random events. The arrival of customers in a queue can be modeled by a Poisson process. In this case, the two principal random variables to consider are:

– the number of customers $N(t)$ arriving in the queue during a length of time t. It is a positive, non-zero, integer-valued random variable: $N(t) \in \mathbb{N}^*$;

– the time T that passes between two consecutive arrivals. We called this the inter-arrival time. It is a real, positive random variable: $T \in \mathbb{R}^+$.

The two random variables $N(t)$ and T are not independent: if the number of customers $N(t)$ arriving in the queue is high, the inter-arrival time is short, and vice versa. Proposition 3.1 explains this relationship between these two random variables characterizing the arrival of customers in a queue.

PROPOSITION 3.1.– *The following three assertions are equivalent:*

1) *The probability that a customer arrives in a queue during an infinitely small interval of time* $\varepsilon > 0$ *equals* $\lambda\varepsilon + o(\varepsilon)$, *with* λ *being a real, positive constant.*

2) *The number of customers* $N(t)$ *arriving in a queue during any interval of time* t *follows a Poisson distribution with averages* λt*:*

$$\mathbb{P}(N(t) = k) = e^{-\lambda t}\frac{(\lambda t)^k}{k!} \qquad [3.1]$$

3) *The inter-arrival* τ *that passes between two consecutive arrivals obeys an exponential distribution with parameter* λ *(average* $1/\lambda$*):*

$$\mathbb{P}(\tau > t) = e^{-\lambda t} \qquad [3.2]$$

PROOF.–

Assertions 2 and 3 were already proven in the first chapter on the definitions of the Poisson process, using the counting measure with the Poisson distribution and the inter-arrivals with the exponential distribution.

The event "a customer arrives in the queue during an infinitely small interval of time ε" from Assertion 1 is equivalent to the event "the inter-arrival τ inferior to ε" deduced from Assertion 3 *whose probability is equal to:*

$$P(\tau < \varepsilon) = 1 - e^{-\lambda\varepsilon}$$

For an infinitely small ε, we can write the limited expansion of $e^{-\lambda\varepsilon}$ near $\varepsilon = 0$: $e^{-\lambda\varepsilon} = 1 - \lambda\varepsilon + o(\varepsilon)$. Thus, we determine that the probability of arrival of a customer in the queue during an infinitely small interval of time ε equals $\lambda\varepsilon + o(\varepsilon)$.

■

3.1.2. *Using the Poisson distribution* $\mathcal{P}(\lambda)$

The question we must now ask is why the number of customers arriving in a queue during an interval of time t practically obeys a Poisson distribution.

We begin with the Bernoulli trial. It was a random experiment with only two possible outcomes; in our case, these outcomes are that a customer arrives at a particular instant or does not. The probability of one of the outcomes equals p and that of the other naturally equals $1-p$. By indefinitely repeating this same trial, we obtain what is called a Bernoulli distribution (or rather the binomial distribution expressed in equation [3.3]) as the distribution of the random variable counting the manifestation of one of these two events during the trials. When the probability p of the event's occurrence is low, we say the event is *rare*.

Let us consider a population of n people capable of independently joining the queue during a unit of time. Each person may join the queue with a probability p or may not join with a probability $1-p$. This is the Bernoulli trial repeated n times for the n people of the population.

The number N of people arriving in the queue follows a Bernoulli distribution $\mathcal{B}(n,p)$:

$$\mathbb{P}(N = k) = \binom{n}{k} p^k (1-p)^{n-k} \tag{3.3}$$

When the number of trials tends towards infinity while the probability p tends towards zero, the Bernoulli distribution in question tends towards a limited distribution that is none other than the Poisson distribution: this is why the Poisson distribution is also called the distribution of rare events.

This Bernoulli distribution tends towards the Poisson distribution for a very high n and $\lambda = pn$

$$\mathcal{P}(\lambda) = \lim_{n\to\infty} \mathcal{B}(n, \lambda/n) \tag{3.4}$$

Since the population that might join the queue is *large*, and the individual choices people make whether to join it or not are independent, we can count the number of customer arrivals in the queue with a random variable following a Poisson distribution. Thus, the three assertions of Proposition 3.1 are applicable (in a general sense) to a queue.

3.1.3. *Exponential distribution of delay times*

Customers arrive in a queue to request some kind of service. Once this service begins, we can suppose that the probability it will be completed after an infinitely small interval of time $\varepsilon > 0$ is $\mu\varepsilon + o(\varepsilon)$. Figure 3.1 represents the transition graph between instants t and $t + \varepsilon$. State 0 indicates a service in progress and state 1 a completed service. The corresponding transition matrix is:

$$\mathbf{P}(\varepsilon) = \begin{pmatrix} 1 - \mu\varepsilon & \mu\varepsilon \\ 0 & 1 \end{pmatrix} + o(\varepsilon) \qquad [3.5]$$

By positing $\mathbf{P}(\varepsilon) = \mathbf{I} + \varepsilon\mathbf{A} + o(\varepsilon)$, we obtain the corresponding infinitesimal generator:

$$\mathbf{A} = \begin{pmatrix} -\mu & \mu \\ 0 & 0 \end{pmatrix} \qquad [3.6]$$

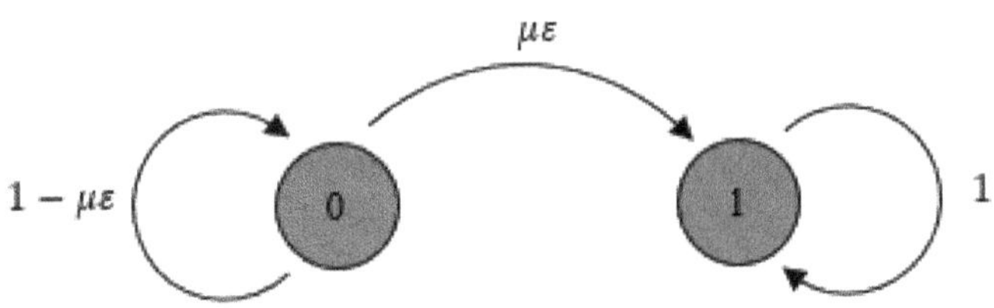

Figure 3.1. *Transition graph of the exponential distribution*

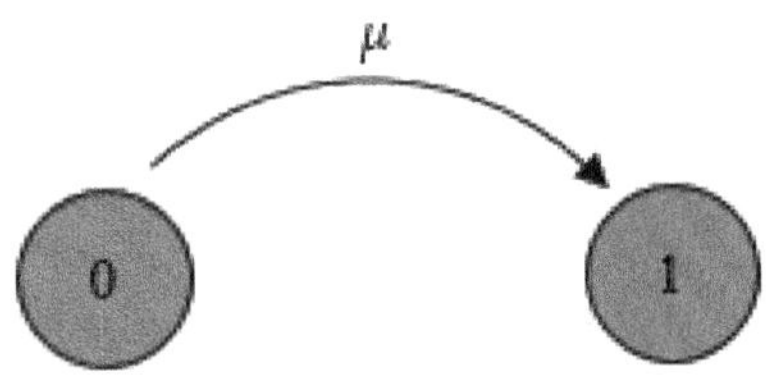

Figure 3.2. *Infinitesimal generator of the exponential distribution*

At the beginning, the service is not yet interrupted (state 0); therefore, the initial distribution is $\boldsymbol{\pi}(0) = (1, 0)$. The differential equation $\dot{\boldsymbol{\pi}}(t) = \boldsymbol{\pi}(t)\mathbf{A}$ is then written as in equation [3.8]:

$$\begin{cases} \dot{\pi}_0(t) = -\mu\pi_0(t) \\ \dot{\pi}_1(t) = \mu\pi_0(t) \\ \pi_0(0) = 1 \\ \pi_1(0) = 0 \end{cases} \qquad [3.7]$$

The solution to this equation is:

$$\begin{cases} \pi_0(t) = e^{-\mu t} \\ \pi_1(t) = 1 - e^{-\mu t} \end{cases} \qquad [3.8]$$

The probability that the service will still be in progress at time t is therefore $e^{-\mu t}$. From this, we can deduce that the length of time T for this service follows an exponential distribution with parameter μ.

We can also say that the exponential distribution is a particular case of the Markov chain.

3.2. Queueing systems

Customers arrive in a queueing system and request a certain service. The instants of arrival and the durations of service are generally random quantities. We model these arrivals with punctual processes. If a reception node is free, the arriving customer goes directly to this node where they are served; otherwise, they join the end of the queue where customers line up to wait.

As shown in Figure 3.3, a queueing system therefore includes a service area with one or more service stations set up in parallel, and a waiting area in which a queue might form.

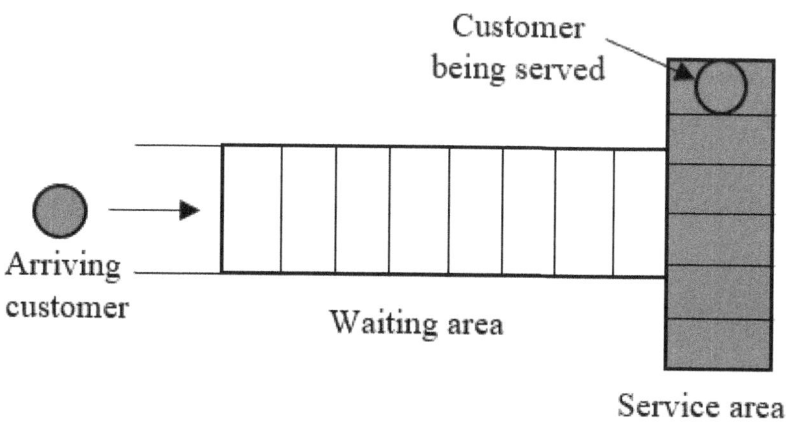

Figure 3.3. *Queueing system*

3.2.1. *Notation for queueing systems*

From the previous description, a queueing system can be characterized by the following parameters:

– the stochastic nature of the arrival process, which is defined by distribution A of the intervals separating two consecutive arrivals;

– distribution B of the duration of service requested by customers;

– the number n of service nodes that are set up in parallel;

– the capacity N of the system (in service + queue). If $N < \infty$, the queue cannot be longer than $N - n$ units. When this capacity is reached, the subsequent customers who approach the system are unable to enter it;

– the total population K of customers who can access the queue;

– the discipline of service Z, which dictates the organization of customers in the waiting area: first in first out (FIFO), last in first out (LIFO), processor sharing (PS), random (R), etc.

For the notation for queueing systems, we have recourse to a symbolic notation containing six symbols in the order $A/B/n/N/K/Z$, known as the Kendall[1] notation. The last three parameters N, K and Z are generally omitted with an infinite default value for N and K, and FIFO for Z.

To specify distributions A and B, we adopt the following convention:

– M: a distribution that satisfies the Markov property, that is, the Poisson process for arrivals, an exponential distribution for service times;

– E_k: an Erlang distribution with parameter k;

– H_k: a hyper-exponential distribution of order k;

– G: a general distribution (we know nothing of its characteristics);

– GI: an independent general distribution;

– D: a deterministic case (the variable only allows for one value).

EXAMPLE.– The notation M/D/1/4 defines a queueing system including one service node for which the longest delay time equals $4 - 1 = 3$. The arrival process is a Poisson process and the duration of service is constant.

1 David George Kendall (1918–2007), British mathematician, specialist in statistics.

3.2.2. *Little distributions*

Little's[2] laws are equations that relate delay times and the number of customers waiting in a queueing system. In everything that follows, when we speak of queueing systems, we refer to the queue itself and the service nodes.

For any queueing system G/G/n/N, we denote by:

– N_S: the number of customers in the system;

– N_q: the number of customers waiting in the queue;

– T_S: the customer holding time in the system, that is, the time spent by the customer from their arrival until the service is completed;

– T_q: the customer holding time in the queue, that is, the delay time before the customer is served;

– λ: the intensity of the arrival process of customers wishing to join the system;

– λ_e: the intensity of the effective arrival process of customers in the system;

– μ: the inverse of the average duration of service requested by customers.

We speak of the arrival process of customers wishing to join the system and the effective arrival process of customers in the case where the capacity of the system is finite. Certain customers wishing to enter risk not being allowed in when the capacity N of the system has been reached and we have $\lambda_e \leq \lambda$. In the case where the system is unlimited, we have $\lambda_e = \lambda$.

We assume that the stationary distributions of the random variables N_S, N_q, T_S and T_q exist, and that the random variable of the duration of service is independent of the delay time T_q of the customer (i.e. we can write $E(T_S) = E(T_q) + 1/\mu$):

PROPOSITION 3.2. (Little theorem).– *The equations that relate delay times and the number of customers in the system are:*

$$\begin{cases} E(N_S) = \lambda_e E(T_S) \\ E(N_q) = \lambda_e E(T_q) \end{cases} \qquad [3.9]$$

PROOF.–

We only offer an intuitive proof. A more detailed proof for the case of an M/M/1 queue is given in section 3.3.2.

2 John Dutton Conant Little (1928–), American professor specializing in operational research.

Let us consider a customer who remains for a time T_s in the system. At the instant when this same customer leaves the system, they have, on average, $\lambda_e T_s$ customers behind them. Therefore, the average number $E(N_s)$ of customers in the system is equal to $E(\lambda_e T_s)$. ■

3.2.3. *Offered traffic*

Even though we do not go into detail about traffic until Chapter 6, we can simply address the one notion of usage rates on resources .

DEFINITION 3.1.– *The percentage of time during which a resource is busy is called the offered traffic .*

The unit of measure for traffic is the *erlang* which we denote Erl in everything that follows.

EXAMPLE.– A telephone line that is busy 100% of the time has a traffic of 1 Erl. Typically, a residential landline has a traffic of 70 mErl, an industrial line 150 mErl and a cellular phone 25 mErl.

For the case of a telephone line, we have the following:

– D: the duration of the observation period;

– d_k: the duration of the k-th call;

– $\bar{d}$: the average duration of a call;

– N_c: the number of calls during an observation period D;

– ρ: the offered traffic in Erl.

The offered traffic is thus:

$$\rho = \frac{1}{D} \times \sum_{k=1}^{N_c} d_k = \frac{N_c \times \bar{d}}{D} \quad [3.10]$$

3.3. M/M/1 queue

From the definition of the notation M/M/1, we have the following hypotheses:

– customers arrive according to a Poisson process with parameter $\lambda > 0$;

– the duration of service follows an exponential distribution with parameter $\mu > 0$;

– the system only has one service node;

– the queue can receive a certain number of customers.

This M/M/1 queue is a particular case of the birth and death process in that the exponential distribution of the duration of service is also a particular case of a Markov chain. The parameter λ_k for the probability that a customer will arrive is independent of the number k of customers in the system; it is always equal to λ, which is the intensity of the Possion process according to Proposition 3.1. Likewise, the parameter μ_k for the probability that a customer will leave the system is also independent of k; it is always equal to μ since the customer will leave once the service is complete. We thus have:

$$\begin{cases} \lambda_k = \lambda & (\forall k \geq 0) \\ \mu_k = \mu & (\forall k \geq 1) \end{cases} \qquad [3.11]$$

3.3.1. *Stationary distribution*

PROPOSITION 3.3.– *Let there be an M/M/1 system with parameters* $\lambda > 0$ *and* $\mu > 0$ *such that* $\lambda < \mu$. *Thus, the number of customers present in the system, in a steady state, follows a geometric distribution with parameter* $\rho := \lambda/\mu$, *that is:*

$$\pi_k = (1-\rho)\rho^k, k = 0,1,2,\ldots \qquad [3.12]$$

PROOF.–

We can demonstrate that the stationary distribution π is a vector with an infinite number of coordinates forming a geometric progression. Equation [2.36] can be written as:

$$\pi_k = \left(\frac{\lambda}{\mu}\right)^k \pi_0, \qquad k = 1,2,\ldots$$

$$\frac{1}{\pi_0} = \sum_{k=0}^{\infty} \left(\frac{\lambda}{\mu}\right)^k = \frac{1}{1-\frac{\lambda}{\mu}}$$

Then, denoting by $\rho = \lambda/\mu$: $\pi_k = \rho^k \pi_0$ et $\pi_0 = (1-\rho)$:

$$\pi_k = (1-\rho)\rho^k, \qquad k = 0,1,2,\ldots$$

In the case where $\rho > 1$, an average of λ customers arrive per unit of time, while the service node, working constantly, can only help μ customers (the average duration of service per customer is $1/\mu$). The number of customers waiting will certainly tend towards infinity and $\sum_{k=0}^{\infty}\left(\frac{\lambda}{\mu}\right)^k$ will diverge.

When $\rho = 1$, equation [3.12] shows that all π_k are zero, which is impossible; the stationary distribution therefore does not exist in this case.

When $\rho < 1$, equation [3.12] defines a vector π with all its strictly positive coordinates (infinite in number). ■

In particular, the probability that the service node will be busy is $1 - \pi_0 = \rho$.

According to the ergodic theorem, the probability π_k corresponds to the percentage of system observation time during which the system contains k customers, when this time tends towards infinity. The fraction of time during which the service node is busy is, in a steady state, equal to $1 - \pi_0 = \rho$. This rate of node activity is, again, called the offered traffic.

In the M/M/1 queue, the offered traffic (equation [3.10]) of the service node is thus equal to:

$$\rho = \frac{\lambda}{\mu} \qquad [3.13]$$

3.3.2. *Characteristics of the M/M/1 queue*

According to Proposition 3.3, which describes the geometric distribution followed by a number N_s of customers in a system, the average and variance are expressed by:

$$E(N_s) = \frac{\rho}{1-\rho} \qquad [3.14]$$

$$\sigma^2(N_s) = \frac{\rho}{(1-\rho)^2} \qquad [3.15]$$

Little's law only tells us about the average time a customer spends in the system, but it is important for us to know how to determine the specific amount of time spent in the system.

PROPOSITION 3.4.– *Let there be an M/M/1 queue with parameters λ and μ with $\lambda < \mu$. In a steady state, the random variable T_s (delay time + service) follows an exponential distribution with parameter $\mu - \lambda$:*

$$\mathbb{P}\,(T_s > t) \;=\; e^{-(\mu\,-\,\lambda)t} \qquad [3.16]$$

PROOF.–

Let us seek the probability distribution of the holding time T_s that a customer spends waiting in the queue. It is natural to think that it is a function of the number of customers already present when the customer arrives.

Let Y_{i-1} be the even value "there are already $i-1$ customers in the system when another customer arrives". Using the law of total probabilities, we have:

$$\mathbb{P}\,(T_s > t) \;= \sum_{i=1}^{+\infty} \mathbb{P}\,(T_s > t|Y_{i-1})\,.\,\mathbb{P}(Y_{i-1})$$

We begin with $i = 1$, since for $i = 0$, there is no waiting and the arriving customer can immediately go up to the service node.

Now, on the one hand, the random variable $(T_s > t|Y_{i-1})$ indicates a holding time that is greater than t given that there are already $i-1$ customers in the system; it is the equivalent to the sum of i independent and identically distributed random variables (i including the customer himself) following an exponential distribution with parameter μ. According to equation [1.9], it is a random variable that follows an Erlang $\mathcal{E}(i,\mu)$ distribution whose density function is equal to:

$$f(x) = \frac{\mu^i x^{i-1}}{(i-1)!} e^{-\mu x}$$

On the other hand, we also have $\mathbb{P}(Y_{i-1}) = \pi_{i-1} = (1-\rho)\rho^{i-1}$ for every $i \geq 1$.

Thus,

$$\mathbb{P}\,(T_s > t) = \sum_{i=1}^{+\infty}\left((1-\rho)\rho^{i-1}\int_t^{+\infty}\frac{\mu^i x^{i-1}}{(i-1)!}e^{-\mu x}dx\right)$$

$$= (1-\rho)\sum_{i=1}^{+\infty}\left(\rho^{i-1}\int_t^{+\infty}\frac{\mu^i x^{i-1}}{(i-1)!}e^{-\mu x}dx\right)$$

$$= (1-\rho)\int_t^{+\infty} e^{-\mu x}\sum_{i=1}^{+\infty}\left(\rho^{n-1}\frac{\mu^i x^{i-1}}{(i-1)!}\right)dx$$

$$= (1-\rho)\int_t^{+\infty} \mu e^{-\mu x}\sum_{i=1}^{+\infty}\frac{(\rho\mu x)^{i-1}}{(i-1)!}dx$$

$$= (1-\rho)\int_t^{+\infty} \mu e^{-\mu x}e^{\rho\mu x}dx = (1-\rho)\int_t^{+\infty} \mu e^{-\mu x}e^{\lambda x}dx$$

$$= (1-\rho)\int_t^{+\infty} \mu e^{(\lambda-\mu)x}dx = (1-\rho)\mu\frac{e^{(\lambda-\mu)t}}{\mu-\lambda}$$

$$= e^{-(\mu-\lambda)t}$$

We indeed arrive at the complementary cumulative distribution function of an exponential distribution with parameter $\mu - \lambda$. ■

The average customer holding time in the system is therefore:

$$E(T_s) = \frac{1}{\mu-\lambda} \qquad [3.17]$$

The proof of the Little theorem for an M/M/1 queue can be easily reached using equations [3.14] and [3.17]:

$$E(N_s) = \frac{\rho}{1-\rho} = \frac{\lambda/\mu}{1-\lambda/\mu} = \frac{\lambda}{\mu-\lambda} = \lambda.E(T_s) \qquad [3.18]$$

3.3.3. *Introducing a factor of impatience*

Let us suppose that a customer who arrives in the queue has the option of leaving the system immediately without receiving the service, depending on the number of customers in the queue before him.

Let us posit, for example, an impatience probability $k/(k+1)$ when k customers are present in the system. If the system is free, this probability equals 0, the customer immediately joins the queue; if k is high, the impatient customer does not dare join the queue, and the impatience probability tends towards 1.

This system can again be modeled by a birth and death process such that the parameter λ_k of the probability of entering the system is:

$$\lambda_k = \lambda\left(1 - \frac{k}{k+1}\right) = \frac{\lambda}{k+1} \qquad (k \geq 0) \qquad [3.19]$$

PROPOSITION 3.5.– *The stationary distribution e corresponding to this type of queue with an impatience factor follows a Poisson distribution with parameter* $\rho = \lambda/\mu$:

$$\pi_k = e^{-\rho}\frac{\rho^k}{k!}, \qquad k \geq 0 \qquad [3.20]$$

PROOF.–

Let us calculate the stationary distribution from equation [2.36]:

$$\pi_k = \frac{\lambda_0\lambda_1 \dots \lambda_{k-1}}{\mu_1\mu_2 \dots \mu_k}\pi_0\,, \qquad k = 1,2, \dots, n$$

$$= \frac{1}{\mu^k}\cdot\frac{\lambda}{1}\frac{\lambda}{2}\cdots\frac{\lambda}{k}\pi_0 = \frac{\lambda^k}{k!\,\mu^k}\pi_0 = \frac{\rho^k}{k!}\pi_0$$

Solving for π_0 gives $\pi_0 = e^{-\rho}$ since $\sum_{k=0}^{+\infty}\pi_k = \sum_{k=0}^{+\infty}\frac{\rho^k}{k!}\pi_0 = 1$.

Thus, $\pi_k = e^{-\rho}\frac{\rho^k}{k!}$ for any $k \geq 1$ and $\pi_0 = e^{-\rho}$. ■

3.4. M/M/∞ queue

From the definition of the notation M/M/∞, we have the following hypotheses:

– customers arrive according to a Poisson process with parameter $\lambda > 0$;

– the duration of service follows an exponential distribution with parameter $\mu > 0$;

– the system has an infinite number of service nodes. No queue will form because every customer who arrives will go directly to a service node.

This system has theoretical value because it allows an approximate study of a queueing system including a large number of service nodes.

This system can be modeled by a birth and death process such that parameters λ_k and μ_k of the probabilities of joining and leaving the system are:

$$\begin{cases} \lambda_k = \lambda & (k \geq 0) \\ \mu_k = k\mu & (k \geq 1) \end{cases} \qquad [3.21]$$

We have k customers served by the service nodes in the queueing system. Each customer has a probability $\mu\varepsilon + o(\varepsilon)$ of leaving the system. The probability that a customer among the k will leave therefore equals $k\mu\varepsilon + o(\varepsilon)$; thus, $\mu_k = k\mu$.

PROPOSITION 3.6.– *The number of customers present in the M/M/∞ system in a steady state follows a Poisson distribution with parameter* $\rho = \lambda/\mu$:

$$\pi_k = e^{-\rho}\frac{\rho^k}{k!} \qquad (k = 0,1,2,\dots) \qquad [3.22]$$

PROOF.–

We primarily use equation [2.36]. The demonstration is identical to that of Proposition 3.5, except that k is with parameter μ, not λ. ■

We deduce that the number N_s of customers being served has as the average and variance (from the expectation and the variance of a random variable following a Poisson distribution):

$$E(N_s) = \sigma^2(N_s) = \rho = \frac{\lambda}{\mu} \qquad [3.23]$$

3.5. M/M/n/n queue

From the definition of the notation M/M/*n*/*n*, we obtain the following hypotheses:

– customers arrive according to a Poisson process with parameter $\lambda > 0$;

– the duration of service follows an exponential distribution with parameter $\mu > 0$;

– the system has exactly n service nodes set up in parallel;

– there will be no queue since the system has a capacity of n customers and these are the n customers being served at the nodes.

This system can be modeled by a birth and death process such that parameters λ_k and μ_k of the probabilities for entering and leaving the system are:

$$\begin{cases} \lambda_k = \lambda & (0 \leq k < n) \\ \mu_k = k\mu & (0 < k \leq n) \end{cases} \quad [3.24]$$

3.5.1. *Stationary distribution*

PROPOSITION 3.7.– *The stationary distribution of the number of customers present in the M/M/n/n system in a steady state is equal to:*

$$\pi_k = \frac{\frac{\rho^k}{k!}}{1 + \rho + \frac{\rho^2}{2!} + \cdots + \frac{\rho^n}{n!}} \qquad (0 \leq k \leq n) \quad [3.25]$$

where $\rho = \lambda/\mu$.

PROOF.–

Provided it can be modeled by a birth and death process, and as with the other queueing systems we have seen up to this point, the calculation of the stationary distribution is performed using equation [2.36]. We leave the reader to continue with the demonstration. ■

3.5.2. *Erlang-B formula*

Using the ergodic theorem, probability π_n is also the fraction of time during which the n service nodes are busy. Once all n nodes are busy, the customer who arrives can no longer enter the system and will be immediately rejected. The probability π_n can thus be interpreted as a loss probability. This notion is very important in telephony and teletraffic .

By writing π_n as $E_{1,n}(\rho)$, we obtain the Erlang loss formula, called the first Erlang formula or the *Erlang-B formula*:

$$\pi_n = E_{1,n}(\rho) = \frac{\frac{\rho^n}{n!}}{1 + \rho + \frac{\rho^2}{2!} + \cdots + \frac{\rho^n}{n!}} \quad [3.26]$$

3.5.3. *Characteristics of the M/M/n/n queue*

We only give the average number $E(N_s)$ of customers in the queueing system here. The average holding time in the system can then be deduced from the Little formulas.

PROPOSITION 3.8.– *In a steady state, the number N_s of customers in the M/M/n/n system is on average:*

$$E(N_s) = \rho.(1 - E_{1,n}(\rho)) \qquad [3.27]$$

PROOF.–

By using the stationary distribution in equation [3.25], we have:

$$\begin{aligned} E(N_s) &= \sum_{k=0}^{n} k\pi_k = \sum_{k=1}^{n} k\pi_k \\ &= \sum_{k=1}^{n} \rho\pi_{k-1} \qquad \text{car } \pi_k = \frac{\rho^k}{k!}\pi_0 \\ &= \rho\sum_{k=0}^{n-1} \pi_k \\ &= \rho(1 - \pi_n) \end{aligned}$$

■

3.6. M/M/n queue

From the definition of the notation M/M/n, we obtain the following hypotheses:

– customers arrive according to a Poisson process with parameter $\lambda > 0$;

– the duration of service follows an exponential distribution with parameter $\mu > 0$;

– the system has exactly n service nodes set up in parallel;

– the system can receive a certain number of customers.

This system can be modeled by a birth and death process such that parameters λ_k and μ_k of the probabilities of entering and leaving the system are:

$$\begin{cases} \lambda_k = \lambda & (0 \leq k) \\ \mu_k = k\mu & (0 < k < n) \\ \mu_k = n\mu & (n \leq k) \end{cases} \qquad [3.28]$$

Indeed, when $k \geq n$, the n customers are served at the nodes and $k-n$ customers can be found in the waiting area. The probability that a customer will leave is calculated on the basis of n customers being served.

3.6.1. *Stationary distribution*

PROPOSITION 3.9.– *The stationary distribution of the number of customers present in the M/M/n system in a steady state is equal to:*

$$\pi_k = \frac{\frac{\rho^k}{k!}}{\frac{\rho^n}{n!}\cdot\frac{n}{n-\rho}+\sum_{i=0}^{n-1}\frac{\rho^i}{i!}} \quad (\text{for } k \leq n)$$

$$\pi_k = \frac{\frac{\rho^k}{n!\,n^{k-n}}}{\frac{\rho^n}{n!}\cdot\frac{n}{n-\rho}+\sum_{i=0}^{n-1}\frac{\rho^i}{i!}} \quad (\text{for } k \geq n) \qquad [3.29]$$

PROOF.–

Using equation [2.36], we can write:

– for $k \leq n$:

$$\pi_k = \frac{\lambda_0}{\mu_1}\frac{\lambda_1}{\mu_2}\cdots\frac{\lambda_{k-1}}{\mu_k}\pi_0 = \frac{\lambda}{\mu}\frac{\lambda}{2\mu}\cdots\frac{\lambda}{k\mu}\pi_0 = \frac{\rho^k}{k!}\pi_0$$

– for $k > n$:

$$\pi_k = \frac{\lambda_0}{\mu_1}\frac{\lambda_1}{\mu_2}\cdots\frac{\lambda_{k-1}}{\mu_k}\pi_0 = \frac{\rho^n}{n!}\left(\frac{\lambda}{n\mu}\right)^{k-n}\pi_0 = \frac{\rho^k}{n!\,n^{k-n}}\pi_0$$

We can deduce the expression for π_0:

$$\frac{1}{\pi_0} = \left(1+\rho+\frac{\rho^2}{2!}+\cdots+\frac{\rho^n}{n!}\right)+\left(\frac{\rho^{n+1}}{n!\,n}+\frac{\rho^{n+2}}{n!\,n^2}+\cdots\right)$$

$$\frac{1}{\pi_0} = \frac{\rho^n}{n!}\sum_{i=0}^{+\infty}\frac{\rho^i}{n^i}+\sum_{i=0}^{n-1}\frac{i}{i!} = \frac{\rho^n}{n!}\cdot\frac{n}{n-\rho}+\sum_{i=0}^{n-1}\frac{\rho^i}{i!}$$

■

3.6.2. *Erlang-C formula*

Using the ergodic theorem, the probability $\sum_{k=n}^{+\infty} \pi_k$ is also the fraction of time during which the n service nodes are busy. Once all n nodes are busy, the customer arriving in the system must wait. The probability $\sum_{k=n}^{+\infty} \pi_k$ can thus be interpreted as a probability of waiting. This notion is also very important in telephony and teletraffic .

By writing $\sum_{k=n}^{+\infty} \pi_k$ as $E_{2,n}(\rho)$, we obtain the Erlang waiting formula, called the second Erlang formula or *the Erlang-C formula*:

$$E_{2,n}(\rho) = \frac{\frac{\rho^n}{n!}\frac{n}{n-\rho}}{\frac{\rho^n}{n!}\frac{n}{n-\rho} + \sum_{i=0}^{n-1}\frac{\rho^i}{i!}} \quad [3.30]$$

3.6.3. *Characteristics of the M/M/n queue*

The average number of customers in the M/M/n queueing system is expressed by:

$$E(N_s) = \rho + \pi_0 \frac{\rho^{n+1}}{(n-1)!\,(n-\rho)^2} \quad [3.31]$$

The delay time T_f follows an exponential distribution with parameter $n\mu - \lambda$ weighted at 0:

$$\mathbb{P}(T_f > t) = \frac{n\rho^n}{n!\,(n-\rho)} e^{-(n\mu-\lambda)} \quad [3.32]$$

The average delay time $E(T_f)$ is equal to:

$$E(T_f) = \pi_0 \frac{n\rho^n}{n!\,(n-\rho)} \frac{1}{n\mu - \lambda} \quad [3.33]$$

The demonstrations of equations [3.31]–[3.33] are left to the reader to complete using analogies with the proofs of the characteristics of the M/M/1 queue in section 3.3.2.

3.7. M/GI/1 queue

From the definition for the notation M/GI/n, we obtain the following hypotheses:

– customers arrive according to a Poisson process with parameter $\lambda > 0$;

– the different durations of service for the customers are independent and identically distributed in a general (non-specified) distribution;

– the system only has one service node;

– the system can receive a certain number of customers.

3.7.1. *Stationary distribution*

The second hypothesis does not allow us to use the birth and death process used in the preceding sections.

PROPOSITION 3.10. (Pollaczek[3]–Khinchine[4] formula).– *The generator function of the stationary distribution of the number of customers in a steady state M/GI/1 queueing system is expressed by:*

$$G(z) = (1-\rho).\frac{\mathcal{L}_S(\lambda(1-z))}{1-\frac{\mathcal{L}_S(\lambda(1-z))-1}{z-1}} \qquad [3.34]$$

where $\mathcal{L}_S(\lambda(1-z))$ designates the Laplace[5] transform for the duration of service.

PROOF.–

We use F to denote the cumulative distribution function of the duration of service and f for its density function.

At instant t, we use N_t to denote the number of customers present in the system and D_t for the duration of service already provided to the customer who is being served. N_t is an integer-valued random variable, and D_t is a real positive random variable.

We only study the process at instants $t_1, t_2, \dots, t_k$, etc., which are the instants when customers 1, 2, …, k, … leave. To simplify the notations, let us posit

3 Félix Pollaczek (1892–1981), Franco-Austrian engineer and mathematician.

4 Aleksandr Yakovlevich Khinchin (1894–1959), Soviet mathematician.

5 Pierre-Simon de Laplace (1749–1827), French mathematician, astronomer, physicist and politician.

$N_k = N_{t_k^+}$, the number of customers present in the system just after the departure of the k-th customer.

If we use $N(t)$ to denote the counting measure up to instant t of the process of arrival of customers in the queueing system, then the number of customers Y_{k+1} who have arrived during the service of the $(k+1)$-th customer is equal to $Y_{k+1} = N(t_{k+1}) - N(t_k)$. The random variables Y_n are independent and identically distributed given that the intervals (t_i, t_{i+1}) are disjointed and the arrival process is a Poisson process. The recurrence equations that relate the sequences N_k to each other are therefore:

$$N_{k+1} = \begin{cases} N_k + Y_{k+1} - 1 & si\ N_k \geq 1 \\ Y_{k+1} & si\ N_k = 0 \end{cases}$$

These equations allow us to define the transitions in state of a homogeneous Markov chain $\{N_k\}_{k\in\mathbb{N}}$.

The transition probabilities of the Markov chain thus defined are determined by the distribution of random variables Y_{k+1}: $a_s = \mathbb{P}(Y_{k+1} = s)$. The Markov chain is also aperiodic and irreducible given that $a_s > 0$ for any $s \in \mathbb{N}$. Conditioning by the duration of service and using the properties of the Poisson process, we have:

$$a_s = \int_0^{+\infty} \frac{(\lambda t)^s}{s!} e^{-\lambda t}.f(t)dt$$

We use $G_A(z) = E(z^Y) = \sum_{s=0}^{+\infty} a_s z^s$ to denote the generator function associated with Y_{k+1}.

The transition probabilities (P_{ij}) of the Markov chain are:

– $P_{ii} = a_1$ for $i \geq 1$ because $N_{k+1} = N_k \geq 1$ for $Y_{k+1} = 1$, that is, a probability of a_1;

– $P_{00} = a_0$ because $N_{k+1} = N_k = 0$ for $Y_{k+1} = 0$, that is, a probability of a_0;

– $P_{ij} = a_{j+1-i}$ for $i \geq 1$ and $j > i$ because $N_{k+1} - N_k = Y_{k+1} - 1 = j - i$, $(N_k \geq 1)$, or rather $Y_{k+1} = j + 1 - i$, that is, a probability of a_{j+1-i};

– $P_{ij} = a_0$ for $i \geq 1$ and $j = i - 1$ because $N_{k+1} - N_k = Y_{k+1} - 1 = j - i = -1$, $(N_k \geq 1)$, in the case where $Y_{k+1} = 0$, that is, a probability of a_0;

– $P_{ij} = a_0$ for $i \geq 1$ and $j < i - 1$ because $N_{k+1} - N_k = Y_{k+1} - 1 < -1$, and $Y_{k+1} < 0$ has a probability of zero;

– $P_{0j} = a_j$ for $i \geq 1$ because $N_{k+1} = Y_{k+1} = j$, that is, a probability of a_j.

The Chapman–Kolmogorov equations (see section 2.2.1.3) that correspond to the stationary distribution $\boldsymbol{\pi}$ can be written as:

$$\pi_k = a_k\pi_0 + \sum_{j=1}^{k+1} \pi_j a_{k+1-j}, \qquad k \in \mathbb{N}$$

To resolve these equations, we make use of the generator series $G(z) = \sum_{k=0}^{+\infty} \pi_k z^k$.

We have:

$$\sum_{k=0}^{+\infty} \pi_k z^k = \sum_{k=0}^{+\infty} a_k \pi_0 z^k + \sum_{k=0}^{+\infty}\sum_{j=1}^{k+1} \pi_j a_{k+1-j} z^k, \qquad k \in \mathbb{N}$$

$$G(z) = \pi_0 G_A(z) + \frac{1}{z}\big(G(z) - \pi_0\big)\big(G_A(z)\big), \qquad k \in \mathbb{N}$$

$$G(z) = \pi_0 G_A(z) + \frac{1}{z}(G(z) - \pi_0)(G_A(z)), \qquad k \in \mathbb{N}$$

$$G(z) = \pi_0 \frac{G_A(z)}{1 - \frac{G_A(z) - 1}{z - 1}}$$

By replacing the generator series $G_A(z)$ with the Laplace transform for the duration of service $G_A(z) = \mathcal{L}_S(\lambda(1-z))$, we obtain the stationary distribution of N_k:

$$G(z) = (1-\rho).\frac{\mathcal{L}_S(\lambda(1-z))}{1 - \frac{\mathcal{L}_S(\lambda(1-z)) - 1}{z-1}}$$

Now, to go from N_k to N_t, we simply use the property that the customers arriving see the M/GI/1 queueing system in a steady state. Consequently, the generator function $G(z)$ established in this way is also that of N_t in a steady state. ■

3.7.2. *Characteristics of the M/GI/1 queue*

From the stationary distribution of equation [3.34], we can determine the average number of customers in the queue:

$$E(N_S) = \rho + \frac{\rho^2(1 + C_S^2)}{2(1-\rho)} \qquad [3.35]$$

where $\rho = \lambda. E(S)$, and C_S^2 is the coefficient of variation for the duration of service, $C_S^2 = \sigma^2(S)/(E(S)^2)$.

As a reminder, the expected value of a random variable X with a known generator series $G_X(z)$ of its probability distribution is expressed by $E(X) = G'(1)$.

The average holding time is deduced from the Little formula.

REMARK.–

Using the results from equation [3.35], we can express the average number of customers in the following types of queues:

– M/D/1 queue: the variance of the duration of service is zero, $C_S^2 = 0$; therefore:

$$E(N_s) = \rho + \frac{\rho^2}{2(1-\rho)} = \frac{\rho}{1-\rho}.\left(1 - \frac{\rho}{2}\right)$$

– M/M/1 queue: the variance of the duration of service is $1/\mu^2$ in the case of an exponential distribution with parameter μ, $C_S^2 = 1$; therefore:

$$E(N_s) = \rho + \frac{2\rho^2}{2(1-\rho)} = \frac{\rho}{1-\rho}$$

We once more arrive at the previously obtained result using another approach.

– M/E_k/1 queue: the duration of service follows an Erlang distribution with parameter k and an average of 1, $C_S^2 = 1/k$; therefore:

$$E(N_s) = \rho + \frac{\rho^2\left(1 + \frac{1}{k}\right)}{2(1-\rho)} = \left(\frac{\rho}{1-\rho}\right)\left(1 - \frac{\rho}{2}\left(1 - \frac{1}{k}\right)\right)$$

3.8. Exercises

EXERCISE 3.1.– *Length of a queue*

Provide the proof for the formula $E(N_q) = E(N_s) - \lambda_e/\mu$.

EXERCISE 3.2.– *M/M/1 queue*

Let us consider an M/M/1 queueing system. A customer arrives on average every 10 minutes and the average duration of service is 7 minutes.

1) Calculate the probability that at least two customers are waiting to be served.

2) Calculate the probability that a customer who arrives must wait before being served.

3) Calculate the probability that a customer who arrives will find a queue of n people ahead of them.

EXERCISE 3.3.– *File transfer*

To transfer a file, we use a line at 512 kbps. The file is transferred by blocks of 100,000 characters of 8 bits.

1) What is the average duration $1/\mu$ needed to transfer one block?

We can suppose that the line delivers a Poissonian traffic with a load limited to 60% (of 512 kbps).

2) What is the rate of arrival in blocks /s?

3) Calculate the average holding time for a block in the queue and the response time of the line.

EXERCISE 3.4.– *Gas station*

A gas station has only one gas pump. Cars arrive according to a Poisson process with the rate of 10 cars per hour. The duration of service follows an exponential distribution with an expected value of 2 minutes.

1) Provide the stationary distribution of the number of cars at the station.

2) Determine the average delay time before getting served and the total holding time in the system.

3) What proportion of cars must wait before being able to fill their tank? What proportion of cars must spend more than two minutes at this station?

Let us now suppose that each driver who encounters two cars at the station leaves immediately.

4) Provide the stationary distribution of the number of cars in the station. What is the probability that a car will leave without filling its tank?

5) Determine the average delay time before being served, and the total holding time in the system.

EXERCISE 3.5.– *At the cybercafé*

Customers arrive in a mini cybercafé according to a Poisson process at a rate of three customers per hour. This mini cybercafé only has two stations. Each customer spends an exponential amount of time using one of the stations of this mini cybercafé with an average of 30 minutes. If a customer arrives and both stations are busy, they leave immediately.

1) Calculate the stationary distribution.

2) Calculate the probability that a customer who arrives will not be served.

We now suppose that this mini cybercafé has a waiting area with two chairs. If a customer arrives and both chairs are occupied, they will leave; otherwise, they will wait until a station becomes free.

3) Calculate the stationary distribution.

4) What is the probability that a customer must wait before being served?

5) Calculate the average delay time.

6) What is the number of customers served per hour?

EXERCISE 3.6.– *Comparison of queues (1)*

Let us consider an M/M/2 queue processing customers at a rate μ, and an M/M/1 queue processing customers at a rate 2μ. The rate of arrival of customers in these two types of queues is λ.

1) Calculate the average number $E(N_S)$ of customers, the average delay time $E(T_q)$ and the average holding time $E(T_S)$ spent in the system for each of these two cases.

2) Draw a comparative conclusion about these two cases.

3) For which of these cases do the service nodes have a greater probability of being busy?

EXERCISE 3.7.– *Comparison of queues (2)*

Let us consider two independent queues, one M/M/1 queue whose rate of arrival of customers is $\lambda/2$ and rate of service is μ and another M/M/2 queue whose rate of arrival is λ and rate of service is μ.

1) Calculate the average number $E(N_S)$ of customers, the average delay time $E(T_q)$ and the average holding time $E(T_S)$ in the system for each of these two cases.

2) Draw a comparative conclusion about these two cases.

3) For which of these cases do the service nodes have a greater probability of being busy?

EXERCISE 3.8.– *Telephone booths*

1) Establish the expressions for the average number of customers and the average holding time in an M/M/s queueing system for the cases $s = 1, 2, 3$ service nodes.

Let us consider a telephone booth with an arrival rate of 6 people/hour that follows a Poisson process. The time spent by an individual in the booth follows an exponential distribution with an average of 5 minutes.

2) What is the average holding time in the booth for each individual?

3) What is the average number of people in the system?

4) What is the probability that the total time spent in the booth will be at least 10 minutes?

5) Let us now suppose that the arrival rate goes up to 20 people/hour. We wish to limit the average number of people to less than three, and the average holding time in the booth to less than 10 minutes. How many telephone booths would be needed, at minimum?

EXERCISE 3.9.– *Telephone switch*

Calls arrive in a telephone switch according to a Poisson process with a rate λ. There are s available lines, and the calls have an exponential duration with an average of $1/\mu$. A call that arrives when all lines are busy is refused.

1) Find the stationary distribution.

2) Calculate the probability that a call will be rejected.

EXERCISE 3.10.– *Queues in tandem*

Let there be a system composed of two service nodes in a series without any waiting areas between the two nodes. A customer is stuck at the exit of the first node, and therefore blocks the whole system as long as the customer being helped at the second node has not finished with their service. Customers arrive at the first node according to a Poisson process with intensity λ and the durations of services follow independent exponential distributions with respective parameters μ_1 and μ_2 for the two service nodes in the system.

1) In a steady state, calculate the probability of a blockage of the system.

2) In order to avoid the blockage, which service is it preferable to place first: the faster or the slower one?

4

Product-Form Queueing Networks

Never will you fulfill an expectation.

Johann Wolfgang von Goethe (1749–1832)

We have examined simple queues, but, in practice, we often encounter queueing networks and not simple queues. We here study an important class of queueing networks commonly used in the application of network optimization problems. They are called Jackson[1] networks, from the name of the mathematician who studied them at the end of the 1950s. Later, during the mid-1960s, four specialists of queues, F. Baskett[2], K.M. Chandy[3], R.R. Muntz[4] and F.G. Palacios[5], in an article that appeared in the *Journal of ACM*, generalized Jackson's results by relaxing certain constraints on the distributions of service. These generalized networks are called BCMP networks, following the initials of the four mathematicians.

This chapter is devoted in particular to the notion of queueing networks. It is only an introduction for those who wish to study further. We will not offer many exercises.

1 James Richard Jackson (1924–2011), American mathematician.

2 Forest Baskett (1943–), professor at the University of Texas in Austin.

3 Kanianthra Mani Chandy (1944–), professor of computer science at the California Institute of Technology.

4 Richard Robert Muntz (1941–), professor at the University of California.

5 Fernando G. Palacios, former student at the University of Texas in Austin.

4.1. Jackson networks

4.1.1. *Definition of a Jackson network*

We define a Jackson network of M/M/1 queues.

Let us consider an *open network* of I queues:

– arrivals: in each queue, i customers arrive following a Poisson process with intensity ν_i;

– demand for service: customers in queue i request i.i.d exponential services with an average of $1/\mu_i$, $i = 1, \dots, I$;

– Jackson routing: after being served from queue i, a customer is directed to queue j with a routing probability of p_{ij} and leaves the network with a probability of $p_i = 1 - \sum_{j=1}^{I} p_{ij}$.

Figure 4.1 illustrates Jackson routing.

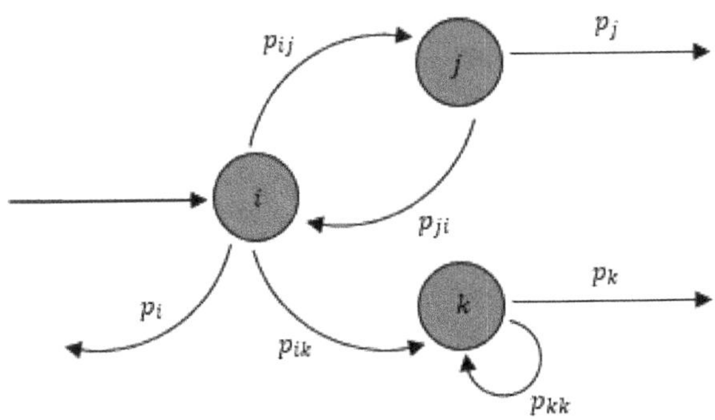

Figure 4.1. *Jackson routing*

DEFINITION 4.1.– *The effective rate of arrival λ_i in queue i is defined by traffic equations*:

$$\lambda_i = \nu_i + \sum_{j=1}^{I} \lambda_j p_{ij} \qquad \forall i \in \{1, \dots, I\} \qquad [4.1]$$

In what follows, we suppose that these equations allow for one unique solution $(\lambda_1, \ldots, \lambda_I)$. Let us also suppose that all customers come from a unique source.

By adding the traffic equations, we obtain a new traffic equation for this unique source:

$$\sum_{i=1}^{I} \nu_i = \sum_{i=1}^{I} \lambda_i p_i \qquad [4.2]$$

The model defined in this way is called a Jackson network of M/M/1 queues.

4.1.2. *Stationary distribution*

Let us use $\rho_i = \lambda_i / \mu_i$ to denote the load factor of the queue i. Let x_i be the number of customers in queue i and $\mathbf{x} = (x_1, \ldots, x_I)$ the state of the network. The state space is thus the set $\mathcal{X} = \mathbb{N}^I$. We denote the capacity of each queue i with $\varphi_i(\mathbf{x})$.

Since the arrival process is Poissonian and the service demands are i.i.d in an exponential distribution, the stochastic process $\{X(t)\}_{t\in\mathbb{R}^+}$ describing the evolution of the number of customers in each queue is a Markovian process.

Let $\mathbf{e}_i$ be the vector whose i –th coordinate is equal to 1 while the others are zero.

Thus, when the network is in state $\mathbf{x}$, the transition rates are zero except for the three following transitions:

– an arrival in queue i bringing the network to state $\mathbf{x} + \mathbf{e}_i$ with the transition rate:

$$q(\mathbf{x}, \mathbf{x} + \mathbf{e}_i) = \nu_i, \qquad \mathbf{x} \in \mathcal{X}$$

– a departure from queue i bringing the network to state $\mathbf{x} - \mathbf{e}_i$ with the transition rate:

$$q(\mathbf{x}, \mathbf{x} - \mathbf{e}_i) = \varphi_i(\mathbf{x}) \mu_i p_i, \qquad \mathbf{x} \in \mathcal{X}, i \geq 1$$

– a displacement from queue i to queue j bringing the network to state $\boldsymbol{x} - \mathbf{e}_i + \mathbf{e}_j$ with the transition rate:

$$q\big(\mathbf{x}, \mathbf{x} - \mathbf{e}_i + \mathbf{e}_j\big) = \varphi_i(\mathbf{x}) \mu_i p_{ij}, \qquad \mathbf{x} \in \mathcal{X}, i, j \geq 1$$

By using the balance equations $\boldsymbol{\pi}(i) \sum_{j \neq i} q(i,j) = \sum_{j \neq i} \boldsymbol{\pi}(j) q(i,j)$, $\forall i \in \mathcal{X}$, the stationary distribution $\boldsymbol{\pi}$ of the number of customers in each queue exists if, and only if, the following global balance equation is satisfied:

$$\begin{aligned} \boldsymbol{\pi}(\mathbf{x}) \sum_{i=1}^{I} (\nu_i + \mu_i) = \sum_{i=1}^{I} \boldsymbol{\pi}(\mathbf{x} - \mathbf{e}_i) \nu_i + \sum_{i=1}^{I} \boldsymbol{\pi}(\mathbf{x} + \mathbf{e}_i) \mu_i p_i \\ + \sum_{i,j} \boldsymbol{\pi}(\mathbf{x} - \mathbf{e}_i + \mathbf{e}_j) \mu_i p_{ij} \end{aligned} \quad [4.3]$$

This equation uses queues with a unitary capacity, that is, queues whose capacity is limited to only one customer, and when another arrives, they are rejected.

If we assimilate a data network to such a network, the capacity of a queue corresponds to the bandwidth allocated to the flow of data on a particular route of the data network. These flows must share the links for common connections with other flows on the same route and also with flows from other routes. The service capacity of a node thus depends on the number of customers at each node. It is thus necessary to study queueing networks with non-unitary capacities that depend rather on the number of customers in each queue.

In this case, we have:

$$\begin{aligned} \boldsymbol{\pi}(\mathbf{x}) \sum_{i=1}^{I} (\nu_i + \varphi_i(\mathbf{x}) \mu_i) \\ = \sum_{i=1}^{I} \boldsymbol{\pi}(\mathbf{x} - \mathbf{e}_i) \nu_i \\ + \sum_{i=1}^{I} \boldsymbol{\pi}(\mathbf{x} + \mathbf{e}_i) \varphi_i(\mathbf{x} + \mathbf{e}_i) \mu_i p_i \\ + \sum_{i,j} \boldsymbol{\pi}(\mathbf{x} - \mathbf{e}_i + \mathbf{e}_j) \varphi_j(\mathbf{x} - \mathbf{e}_i + \mathbf{e}_j) \mu_i p_{ij} \end{aligned} \quad [4.4]$$

PROPOSITION 4.1 (Burke[6] theorem for open queueing networks).– *Provided that the open Jackson network is stable, in a steady state, the departure process is Poissonian.*

PROOF.– Accepted.

6 Paul J. Burke, American engineer, former employee of Bell Telephone Laboratories.

Let us posit $N = (N_1, \cdots, N_n)$, the random vector representing the number of customers in the n nodes of the network. We obtain the following important result.

PROPOSITION 4.2 (Jackson theorem).– *Provided that the queueing network is stable:* $\rho_i < 1\, \forall i$:

Open networks

The stationary probability distribution for the number of customers in the network is factored into a product of marginal probabilities:

$$\mathbb{P}(N_1 = k_1, \dots, N_n = k_n) = \mathbb{P}(N_1 = k_1) \dots \mathbb{P}(N_n = k_n) \qquad [4.5]$$

Each of these marginal probabilities $\mathbb{P}(N_i = k_i)$ is defined by:

$$\mathbb{P}(N_i = k_i) = \frac{b_i \lambda_i^{k_i}}{\mu_i(1) \dots \mu_i(k_i)} \qquad [4.6]$$

with b_i being a normalization constant satisfying:

$$\frac{1}{b_i} = \sum_{k=0}^{\infty} \frac{b_i \lambda_i^k}{\mu_i(1) \dots \mu_i(k)} \qquad [4.7]$$

and $\mu_i(l)$ the service capacity offered by node i when the number of customers present in the node is l; the traffic λ_i entering nodes i is given by the traffic equations defined above.

Closed networks

The distribution of stationary probability can be written as:

$$\begin{aligned} \mathbb{P}(N_1 = k_1, \dots, N_n = k_n) \\ &= B_K \frac{\lambda_1^{k_1}}{\mu_1(1) \dots \mu_1(k_1)} \cdots \frac{\lambda_n^{k_n}}{\mu_n(1) \dots \mu_n(k_n)} \\ &= B_K \prod_{i=1}^{n} \frac{\lambda_i^{k_i}}{\mu_i(1) \dots \mu_i(k_i)} \end{aligned} \qquad [4.8]$$

where B_K is a normalization constant defined by:

$$B_K = \frac{1}{\sum_K \left(\prod_{i=1}^{n} \frac{\lambda_i^{k_i}}{\mu_i(1) \dots \mu_i(k_i)} \right)} \quad [4.9]$$

with $K = k_1 + \cdots + k_n$ representing the finite number of customers in the closed network.

REMARK 4.1.– In the case of a closed network, since the number of customers is finite, the lengths of queues are not independent. Therefore, *stricto sensu*, we do not have a product form since the distribution cannot be factored into the product of marginal probabilities.

4.1.3. *The particular case of the Jackson theorem for open networks*

To simplify the expressions, let us suppose that all the queues in the network are M/M/1, thus $\forall i \ \mu_i(k) = \mu_i \ \forall k$,

$$\frac{1}{b_i} = \sum_{k=0}^{\infty} \frac{\lambda_i^k}{\mu_i^k} = \sum_{k=0}^{\infty} \rho_i^k = \frac{1}{1-\rho_i} \quad [4.10]$$

The marginal distribution is then given by:

$$\mathbb{P}(N_i = k_i) = \rho_i^k (1-\rho_i), \quad i = 1, \dots, n \quad [4.11]$$

From the marginal distribution of queue i, we can calculate the average number of customers in the queue and, using the Little theorem, the average time spent going through node i. Thus, for any given path in the network, it is possible, by adding the different times spent in the nodes of the path, to calculate the average time spent in traveling through this path.

4.1.4. *Generalization of Jackson networks: BCMP networks*

F. Baskett, K.M. Chandy, R.R. Muntz and F.G. Palacios showed that the probability distribution of the number of packets in the network is factored in product form under more general conditions than in the Jackson theorem:

– with a finite number of packet classes k, with each class following an exponential distribution of service with parameters $\{\mu_i^j\}; i = 1, \dots, n; j = 1, \dots, k$ for a network with n nodes and k classes;

– in the case of just one class of packets, in the following queues:

- M/GI/1 with the policy of "processor sharing",
- M/GI/∞,
- M/GI/*s* with the LIFO policy and a "preemptive resume" priority.

We can also impose a technical mathematical constraint on the Laplace transform of the required service.

What we call the policy of "processor sharing" refers to the sharing of a server between customers present in the system. To get a clearer picture, if in the system, at an instant t, there are x customers present, then in the space of one second of time, each customer will receive a service of $1/x$ seconds. Customers are served cyclically. It is therefore clear that this policy is of the "time sharing" type, and the length of time a processor can be allocated to a customer per unit of time varies with respect to the number of customers in the system.

What we call a "preemptive resume" LIFO policy is one in which the precedence is given to the last customer to arrive. The customer who is given the priority will interrupt the customer whose service is in progress. However, the service already allocated to a customer is not lost. The service of the interrupted customer will recommence at the point where the interruption took place.

By way of a comment on these general services, it is clear that they are not very appropriate for the modeling of packet-switched networks. However, the possibility of accounting for numerous classes of customers allows us to model traffic (with exponentially distributed service) of different natures (e.g. X25 packet traffic and datagram traffic).

4.2. Whittle networks

4.2.1. *Definition of a Whittle network*

Whittle[7] networks constitute a particular class of queueing networks, characterized by their balance property.

DEFINITION 4.2.– *Service capacities in a queueing network are called* balanced *if:*

$$\begin{gathered}\varphi_i(\mathbf{x})\varphi_j(\mathbf{x}-\mathbf{e}_i) = \varphi_j(\mathbf{x})\varphi_i(\mathbf{x}-\mathbf{e}_j),\\ i,j = 1,\dots,I:\ x_i > 0, x_j > 0.\end{gathered} \qquad [4.12]$$

7 Peter Whittle (1927–2021), mathematician and statistician from New Zealand.

Let us consider a direct path from state x to state 0, $\langle \mathbf{x}, \mathbf{x}-\mathbf{e}_{i_1}, \ldots, \mathbf{x}-\mathbf{e}_{i_1}-\cdots-\mathbf{e}_{i_{c-1}}, 0\rangle$; this path is of length c, where $c = x_1 + \ldots + x_I$ gives the total number of customers in state **x**.

DEFINITION 4.3.– *We define a balance function denoted by* Φ *as the function expressed by:*

$$\Phi(\mathbf{x}) = \frac{1}{\varphi_{i_1}(\mathbf{x})\varphi_{i_2}(\mathbf{x}-\mathbf{e}_{i_1}) \ldots \varphi_{i_c}(\mathbf{x}-\mathbf{e}_{i_1}-\cdots-\mathbf{e}_{i_{c-1}})} \quad [4.13]$$

The balance property requires that the balance function be independent of the path from state **x** to state **0**.

PROPOSITION 4.3.– Capacities $\varphi_i(\mathbf{x})$ are characterized by the balance function Φ:

$$\varphi_i(\mathbf{x}) = \frac{\Phi(\mathbf{x}-\mathbf{e}_i)}{\Phi(\mathbf{x})}, \qquad i = 1, \ldots, I, \ x_i > 0 \quad [4.14]$$

We say that these capacities are balanced by function Φ.

PROOF.–

We have:

$$\Phi(\mathbf{x}) = \frac{1}{\varphi_{i_1}(\mathbf{x})\varphi_{i_2}(\mathbf{x}-\mathbf{e}_{i_1})\varphi_{i_3}(\mathbf{x}-\mathbf{e}_{i_1}-\mathbf{e}_{i_2}) \ldots \varphi_{i_c}(\mathbf{x}-\mathbf{e}_{i_1}-\cdots-\mathbf{e}_{i_{c-1}})}$$

Therefore, by replacing coordinate **x** by $\mathbf{x}-\mathbf{e}_{i_1}$, we obtain:

$$\Phi(\mathbf{x}-\mathbf{e}_{i_1}) = \frac{1}{\varphi_{i_1}(\mathbf{x}-\mathbf{e}_{i_1})\varphi_{i_2}(\mathbf{x}-\mathbf{e}_{i_1}-\mathbf{e}_{i_2})\varphi_{i_3}(\mathbf{x}-\mathbf{e}_{i_1}-\mathbf{e}_{i_1}-\mathbf{e}_{i_2}) \ldots}$$

Now, according to the definition of balanced capacities:

$$\varphi_i(\mathbf{x})\varphi_j(\mathbf{x}-\mathbf{e}_i) = \varphi_j(\mathbf{x})\varphi_i(\mathbf{x}-\mathbf{e}_j)$$

First, we change i to i_1, j to i_2 and **x** to $\mathbf{x}-\mathbf{e}_{i_1}$:

$$\varphi_{i_1}(\mathbf{x}-\mathbf{e}_{i_1})\varphi_{i_2}(\mathbf{x}-\mathbf{e}_{i_1}-\mathbf{e}_{i_1}) = \varphi_{i_2}(\mathbf{x}-\mathbf{e}_{i_1})\varphi_{i_1}(\mathbf{x}-\mathbf{e}_{i_1}-\mathbf{e}_{i_2})$$

$$\varphi_{i_2}(\mathbf{x}-\mathbf{e}_{i_1}-\mathbf{e}_{i_1}) = \frac{\varphi_{i_2}(\mathbf{x}-\mathbf{e}_{i_1})\varphi_{i_1}(\mathbf{x}-\mathbf{e}_{i_1}-\mathbf{e}_{i_2})}{\varphi_{i_1}(\mathbf{x}-\mathbf{e}_{i_1})}$$

and $\Phi(\mathbf{x}-\mathbf{e}_{i_1})$ becomes:

$$\Phi(\mathbf{x}-\mathbf{e}_{i_1}) = \frac{1}{\varphi_{i_1}(\mathbf{x}-\mathbf{e}_{i_1})\frac{\varphi_{i_2}(\mathbf{x}-\mathbf{e}_{i_1})\varphi_{i_1}(\mathbf{x}-\mathbf{e}_{i_1}-\mathbf{e}_{i_2})}{\varphi_{i_1}(\mathbf{x}-\mathbf{e}_{i_1})}\varphi_{i_3}(\mathbf{x}-\mathbf{e}_{i_1}-\mathbf{e}_{i_1}-\mathbf{e}_{i_2})\dots}$$

$$\Phi(\mathbf{x}-\mathbf{e}_{i_1}) = \frac{1}{\varphi_{i_2}(\mathbf{x}-\mathbf{e}_{i_1})\varphi_{i_1}(\mathbf{x}-\mathbf{e}_{i_1}-\mathbf{e}_{i_2})\varphi_{i_3}(\mathbf{x}-\mathbf{e}_{i_1}-\mathbf{e}_{i_1}-\mathbf{e}_{i_2})\dots}$$

Then, in the definition of balanced capacities, we change i to i_1, j to i_3 and $\mathbf{x}$ to $\mathbf{x}-\mathbf{e}_{i_1}-\mathbf{e}_{i_2}$:

$$\varphi_{i_1}(\mathbf{x}-\mathbf{e}_{i_1}-\mathbf{e}_{i_2})\varphi_{i_3}(\mathbf{x}-\mathbf{e}_{i_1}-\mathbf{e}_{i_2}-\mathbf{e}_{i_1}) = \varphi_{i_3}(\mathbf{x}-\mathbf{e}_{i_1}-\mathbf{e}_{i_2})\varphi_{i_1}(\mathbf{x}-\mathbf{e}_{i_1}-\mathbf{e}_{i_2}-\mathbf{e}_{i_3})$$

$$\varphi_{i_3}(\mathbf{x}-\mathbf{e}_{i_1}-\mathbf{e}_{i_2}-\mathbf{e}_{i_1}) = \frac{\varphi_{i_3}(\mathbf{x}-\mathbf{e}_{i_1}-\mathbf{e}_{i_2})\varphi_{i_1}(\mathbf{x}-\mathbf{e}_{i_1}-\mathbf{e}_{i_2}-\mathbf{e}_{i_3})}{\varphi_{i_1}(\mathbf{x}-\mathbf{e}_{i_1}-\mathbf{e}_{i_2})}$$

and $\Phi(\mathbf{x}-\mathbf{e}_{\mathbf{i}_1})$ becomes:

$$\Phi(\mathbf{x}-\mathbf{e}_{i_1}) = \frac{1}{\varphi_{i_2}(\mathbf{x}-\mathbf{e}_{i_1})\varphi_{i_1}(\mathbf{x}-\mathbf{e}_{i_1}-\mathbf{e}_{i_2})\frac{\varphi_{i_3}(\mathbf{x}-\mathbf{e}_{i_1}-\mathbf{e}_{i_2})\varphi_{i_1}(\mathbf{x}-\mathbf{e}_{i_1}-\mathbf{e}_{i_2}-\mathbf{e}_{i_3})}{\varphi_{i_1}(\mathbf{x}-\mathbf{e}_{i_1}-\mathbf{e}_{i_2})}\dots}$$

$$\Phi(\mathbf{x}-\mathbf{e}_{i_1}) = \frac{1}{\varphi_{i_2}(\mathbf{x}-\mathbf{e}_{i_1})\varphi_{i_3}(\mathbf{x}-\mathbf{e}_{i_1}-\mathbf{e}_{i_2})\varphi_{i_1}(\mathbf{x}-\mathbf{e}_{i_1}-\mathbf{e}_{i_2}-\mathbf{e}_{i_3})\varphi_{i_4}\dots}$$

By iterating the same procedure until i is replaced by i_1, j by i_c and $\mathbf{x}$ by $\mathbf{x}-\mathbf{e}_{i_1}-\mathbf{e}_{i_2}-\dots-\mathbf{e}_{i_{c-1}}$, and recalling that $\mathbf{x}-\mathbf{e}_{i_1}-\mathbf{e}_{i_2}-\dots-\mathbf{e}_{i_c}=\mathbf{0}$, we finally obtain:

$$\Phi(\mathbf{x}-\mathbf{e}_{i_1}) = \frac{1}{\varphi_{i_2}(\mathbf{x}-\mathbf{e}_{i_1})\varphi_{i_3}(\mathbf{x}-\mathbf{e}_{i_1}-\mathbf{e}_{i_2})\dots\varphi_{i_c}(\mathbf{x}-\mathbf{e}_{i_1}-\dots-\mathbf{e}_{i_{c-1}})\varphi_{i_1}(\mathbf{0})\varphi_{i_c}(\mathbf{0})}$$

$$\Phi(\mathbf{x}-\mathbf{e}_{i_1}) = \frac{1}{\varphi_{i_2}(\mathbf{x}-\mathbf{e}_{i_1})\varphi_{i_3}(\mathbf{x}-\mathbf{e}_{i_1}-\mathbf{e}_{i_2})\dots\varphi_{i_c}(\mathbf{x}-\mathbf{e}_{i_1}-\dots-\mathbf{e}_{i_{c-1}})}$$

The first equation is:

$$\Phi(\mathbf{x}) = \frac{1}{\varphi_{i_1}(\mathbf{x})\varphi_{i_2}(\mathbf{x}-\mathbf{e}_{i_1})\varphi_{i_3}(\mathbf{x}-\mathbf{e}_{i_1}-\mathbf{e}_{i_2})\ldots\varphi_{i_c}(\mathbf{x}-\mathbf{e}_{i_1}-\cdots-\mathbf{e}_{i_{c-1}})}$$

By reducing the equation, we get:

$$\frac{\Phi(\mathbf{x})}{\Phi(\mathbf{x}-\mathbf{e}_{i_1})} = \frac{1}{\varphi_{i_1}(\mathbf{x})}$$

which brings us to:

$$\varphi_{i_1}(\mathbf{x}) = \frac{\Phi(\mathbf{x}-\mathbf{e}_{i_1})}{\Phi(\mathbf{x})} \qquad \blacksquare$$

4.2.2. *Stationary distribution*

PROPOSITION 4.4.– *In a stable Whittle network, the stationary distribution of the number of customers in each queue is given by:*

$$\boldsymbol{\pi}(\mathbf{x}) = K_\pi \Phi(\mathbf{x}) \prod_{i=1}^{I} \left(\frac{\lambda_i}{\mu_i}\right)^{x_i}, \qquad \mathbf{x} \in \mathcal{X} \qquad [4.15]$$

where K is a normalization constant as in the definition of $\boldsymbol{\pi}$.

This invariable measure does not depend on the distribution of quantities of required services in each queue, apart from its average. We say that this network has the property of *insensitivity*.

PROOF.– Accepted.

4.2.3. *Properties of a Whittle network*

A Whittle network satisfies the following hypotheses:

– when the network is in state $\mathbf{x}$, the time preceding the movement of just one unit going from node i to node j follows an exponential distribution with parameter $\varphi_j(\mathbf{x})\mu_j p_{ij}$;

– capacities $\varphi_i(\mathbf{x})$ satisfy the balance property.

Consequently, a Jackson network is a Whittle network in which the capacity at each node does not depend on the number of units in the node.

A Whittle process is a Markovian process associated with the evolution of the number of customers in a Whittle network.

The transitions in such a process are given by:

$$q(\mathbf{x},\mathbf{y}) = \begin{cases} \varphi_i(\mathbf{x})p_{ik} & if\ \mathbf{y} = \mathbf{x} + \mathbf{e}_k - \mathbf{e}_i \in \mathcal{X} \\ \pi_j & if\ \mathbf{y} = \mathbf{x} + \mathbf{e}_j \in \mathcal{X} \\ 0 & otherwise \end{cases} \qquad [4.16]$$

4.3. Exercise

A Jackson network is a set of N queues of type G/M/c_p whose service rates are characterized by $\mu_p(n_p) = \mu_p \min(c_p;\ n_p)$, where n_p is the number of people in queue p and c_p the number of servers in parallel. These queues are connected to each other such that at the exit of p, one can:

– join queue q with a probability of $r_{p,q}$;

– leave the system definitively with a probability of $r_{p,N+1}$.

In what follows, let us suppose that $\mu_p(n_p) > 0$ when $n_p > 0$ and $r_{p,p} = 0$ to simplify things. Exogenous arrivals (from outside the system) in queue p form a Poisson process with intensity $\bar{\lambda}_p$. We affirm that vector $\mathbf{X}(t)$ for the number of people in the queue is a Markov chain in continuous time.

1) Draw this system and describe the infinitesimal generator of $\mathbf{X}(t)$.

Let us suppose the system is in a steady state.

2) Provide the balance equations of the queues.

3) Show that if the network has no capture, that is, if any packet has a non-zero probability of leaving the system, then the previous system has one positive and finite solution.

4) Demonstrate the following theorem:

THEOREM 4.1.– *Let there be an irreducible Markov chain in continuous time with an infinitesimal stochastic generator* $\mathbf{Q} = (q_{ij})$. *If we can find a distribution* $\boldsymbol{\pi}$ *and positive numbers* $\tilde{q}_{ij}$ *(for* $i \neq j$*) such that* $\boldsymbol{\pi}(i)\tilde{q}_{ij} = \boldsymbol{\pi}(j)q_{ji}$ *and* $\sum_{j \neq i} \tilde{q}_{ij} = -q_{ii}$, *then* $\boldsymbol{\pi}$ *is the stationary distribution of the chain.*

Let us suppose that each queue is M/M/s and independent of the others.

5) What is the expected stationary distribution for $\mathbf{X}(t)$?

6) Show that it is indeed the stationary distribution of the system (although the chains are not independent).

PART 3

Teletraffic

5

Notion of Teletraffic

When you're poor, the only resource left is to be wise.

Jean-Pierre Florian (1755–1794)

Resources are the most important parts in a telecommunication network. They are generally rare and always need to be shared. This sharing concerns almost all of the components of a network:

– transmission channels, memory, processors;

– transmission components (repeaters, amplifiers);

– switching components (telephone exchanges, routers, couplers);

– control devices (processors, signaling equipment);

– even the resources we think are indivisible (power, bandwidth, frequency spectrums).

This chapter is devoted to the introduction of a notion that is useful in the mathematical study of this kind of sharing, which is based on statistical speculation relating to the number of users who might make use of these components, relative to the available number of them.

5.1. Teletraffic and its objectives

DEFINITION 5.1.– *Teletraffic can be defined as a stochastic process that corresponds to the set of uses (real or fictitious) of resources in a telecommunication network, whatever their cause and independently of whether or not they are connected to complete, effective communication.*

According to this definition, teletraffic has a stochastic nature. It depends on random variables that are themselves dependent on time. We can cite discrete random variables such as the number of requests to use resources, the number of resources that are busy, the length of the messages in need of these resources, etc. There are also continuous random variables such as the length of time the resources remain busy.

The principal objective of teletraffic is to evaluate the risk caused by the sharing of these resources so as to decide on appropriate dimensions, ensuring a better compromise between the cost of resources used and the quality of the service making use of said resources. Indeed, sharing implies a limited number of resources accessible to numerous users; they are only temporarily attributed for the duration of services by their users. The quality of the service demands that the resources be available when a user makes a request in order to meet their needs.

Managing teletraffic uses probability and statistical tools, such as queues, the nature of traffic, practical models, measures and simulations, in order to make predictions and plan out telecommunication networks, such as phone networks or the Internet. The measure of existing traffic must account for the type of service, the spatiotemporal environment and the behavior of users. Simulations are created from established models to describe the sources and resources of traffic processing. These tools help in the provision of reliable services at a reduced cost.

5.2. Definitions

5.2.1. *Measures in teletraffic*

As mentioned in Definition 5.1, teletraffic is a stochastic process. Even though the measures found in teletraffic are random variables that are dependent on time, we generally pose the hypothesis of a stationary stochastic process. The statistical properties of such a process are independent of time, and they will facilitate our analysis of the system being considered. Nevertheless, however helpful it may be, this hypothesis is often false, especially in the case where the observation period of the system is too long.

In teletraffic, we cannot use instantaneous values for the considered measures. They lead to erroneous results since these measures are entirely random and depend on the moment of observation. For example, Internet traffic in a zone may have entirely atypical congestion at around midnight of the last day of the year going into the start of the year; using the teletraffic measures from this moment will lead to a stark over-dimensioning of the system relative to the values of a normal day. An over-dimensioning can impact the profitability of the system.

If we cannot use instantaneous values, we thus have recourse to average values, especially for estimating traffic. However, an average value over a long period of observation will lead to erroneous results in the dimensioning. Indeed, the use of resources during nighttime lulls is very minimal and lowers the average value of usage. This leads to an under-dimensioning of the network, impacting in turn the quality of the service offered to users.

A very precise rule must be considered for the calculation of these average times. Generally, an average is taken from an hour considered characteristic, called the *busy hour*.

DEFINITION 5.2.– *The busy hour is a continuous period of one hour entirely contained within the pertinent interval of time for which the usage of resources or the number of attempts at usage is maximal.*

EXAMPLE.– The busy hour for voice traffic is often situated around the time in the evening when work finishes (6–7 pm in Madagascar), which happens again and again for every day of the week throughout the year. With Internet, the traffic becomes equally high at the start of the evening (8–9 pm in Madagascar).

5.2.2. Sources and resources

DEFINITION 5.3.– *What we call a source is any entity exterior to the system (network) that can use the resources of the system for any service.*

A source can also be called a "user". In other words, it is the element of a population capable of initiating a demand for resources from a system to meet their needs. The number of sources can be finite or infinite depending on the case study.

We can also speak of the "origin" and "destination" of traffic. The origin is the place, which we can identify with the requisite precision, where the source is located; meanwhile the destination is the location of the requested endpoint (if the requested service consists of connecting the source to this endpoint).

DEFINITION 5.4.– *What we call a resource is any group of entities that can be defined materially or conceptually within a system whose usage can be determined without ambiguity.*

A resource unit can be free or busy depending on whether or not it is taken by a source.

5.2.3. *Requests and holding time*

In order to be attributed one or more resources, a user must make a request for this service. In the case of telephony, this request is called a "call attempt", but in a general sense, we will speak of *requests*.

DEFINITION 5.5.– *A request is an individual attempt at obtaining a service from a resource of a particular type.*

In other words, the word "request" designates the activation of a resource, therefore moving it from an inactive state to an active state, or initiating activity. This punctual event without any associated duration may or may not lead to activity depending on the availability of the resources in the system and that of the requested endpoint.

DEFINITION 5.6.– *The holding time is the period of time between when a resource is taken and then freed.*

Activity is therefore the result of a successful request. The request is accepted by the system offering the requested service, and as a result, the necessary resources for the implementation of the service become occupied.

Definition 5.1 of teletraffic mentions fictitious activity. This refers to the activity of estimated traffic before this traffic has actually occurred. Fictitious activity cannot lead to effective activity.

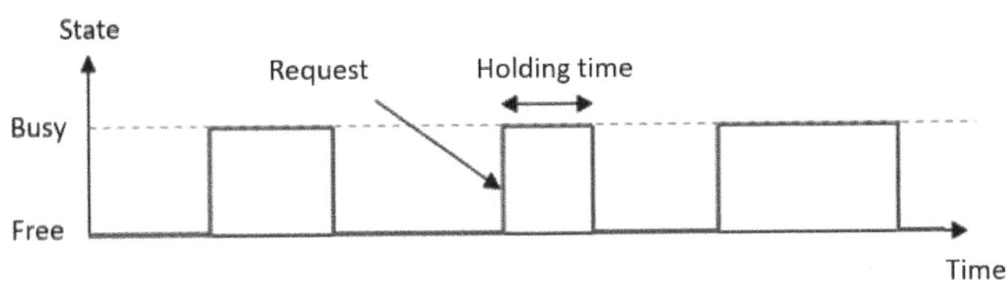

Figure 5.1. *Requests and activity*

5.2.4. *Traffic*

5.2.4.1. *The directional phenomenon of traffic*

DEFINITION 5.7.– *Traffic is a directional phenomenon, in the sense that it relates a source (that has generated a request) and a recipient.*

Once a request is accepted by the system, the resulting activity of resources can give way to a bidirectional exchange of information between the source and the recipient. This exchange says nothing of the direction of traffic, which is defined with respect to the request made by the source.

However, in certain cases, the recipient must in turn also make a request for resources in order to communicate with "the source". This is most often found in the case of bidirectional communications.

In telecommunication, the direction of traffic can be imposed by the type of connection used in the transmission environment.

DEFINITION 5.8.– *What we call departure traffic is the traffic generated by a collection of sources, created in the system in question, independently of the traffic's destination.*

DEFINITION 5.9.– *What we call arrival traffic is the traffic received by a collection of recipients, directed towards the system in question, independently of the origin.*

DEFINITION 5.10.– *What we call internal traffic is departure traffic that is also arrival traffic for the system in question.*

In the case of telecommunication networks, we can also speak of entering traffic and exiting traffic with respect to a node.

DEFINITION 5.11.– *What we call incoming traffic is the traffic coming from the exterior that, whatever the destination, penetrates the network in question.*

DEFINITION 5.12.– *What we call outgoing traffic is the traffic destined for the exterior that, whatever the origin, leaves the network in question.*

These definitions are represented in Figure 5.2.

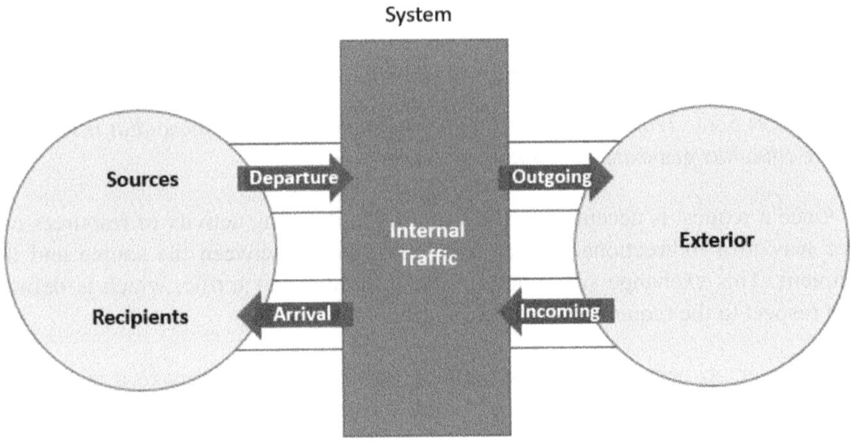

Figure 5.2. *Breakdown of traffic into its different components*

5.2.4.2. *Carried traffic and offered traffic*

The concept of traffic intensity was already evoked in section 3.2.2 where we studied M/M/1 queues. Let us recall Definition 3.1, according to which traffic is the percentage of time that a resource at a given service station in an M/M/1 system is busy. This time, we provide a more general definition.

DEFINITION 5.13.– *The intensity of traffic Y carried by a group of resources is defined by the product of the average frequency* c *of requests leading to said resources being busy and the average* h *of this activity*:

$$Y = c.h \quad [5.1]$$

To simplify the language, we simply say *carried traffic* instead of the intensity of carried traffic.

Traffic is the product of a frequency expressed by the unit of time and a time. It is therefore a measure without dimensions, even though it is expressed in *erlang* (Erl). This unit comes from the name of the Danish mathematician A.K. Erlang, to whom we owe the mathematical theory of teletraffic.

In a system that only has one available resource, carried traffic is thus equal to the proportion of time during which this resource is busy. In other words, it is the probability of finding this resource busy when a user makes a request at a random instant. This reflects Definition 3.1 *set out for the case of an M/M/1 system.* Consequently, the value of traffic carried by one resource cannot surpass 1 Erl.

Let us posit that the total traffic Y carried by a collection of n resources is equal to the sum of traffic Y_i carried individually by each of these resources:

$$Y = \sum_{i=1}^{n} Y_i \qquad [5.2]$$

The traffic carried by a collection of N resources cannot surpass n Erl.

The intensity of the traffic can also be interpreted as the *load* of a system without providing any information about the nature of this load, such as the proportion of long periods of activity or that of short periods of activity.

DEFINITION 5.14.– *What we call offered traffic A is the traffic generated by the collection of requests, whether they are accepted or refused.*

Using c_A to denote the frequency of all the requests without distinguishing whether or not they are accepted and h for the average duration of activity, the amount of offered traffic A is expressed by:

$$A = c_A.h \qquad [5.3]$$

Since $c \leq c_A$, the carried traffic is always inferior or equal to the offered traffic. It does not account for real activity.

REMARK.– To distinguish these two types of traffic, we can also memorize the following notion. Users generate traffic from the requests they make, but this traffic is fictitious, and it is this kind of traffic which is offered to the system: hence its name “offered traffic”. From the system’s perspective, it will handle the traffic offered by users depending on the resources it has. It is possible that all the traffic it is offered cannot be carried due to a lack of resources. The traffic that is taken forward (accepted requests) is called carried traffic, and that which cannot be taken forward will be lost, put on hold or even routed to another system that can process the request.

5.2.4.3. *Temporal variation of traffic*

As mentioned in section 5.2.1, measures in teletraffic, and more specifically the intensity of traffic, are random variables that depend on time. Figure 5.3 shows an example of traffic evolving over time in the case of a cellular phone network. This figure allows us to conclude that traffic depends on several factors:

– the time (working hours, breaktimes, weekends, etc.);

– the place (urban, rural, market, etc.);

– tariff policies (reduced evening rates, for example);

– exceptional circumstances (catastrophes, cultural or sporting events, public holidays, radio games, etc.).

Even though this figure only concerns telephone traffic, the enumerated conclusions are indeed applicable in more general cases of teletraffic. These characteristics also allow us to gauge other systems such as the queues in a post office, for bank tellers, for the stock at stores, for the checkout, etc.

Figure 5.3 represents the traffic carried by a GSM base transceiver station (BTS) in Analakely (a neighborhood in Antananarivo, the capital of Madagascar) over consecutive days of the week. The measurements were made in the month of June 2021, but they remain perfectly representative of the traffic generated principally by cellular phone users. The study of the different curves shows the direct relationship between social habits and the planning of a cellular network.

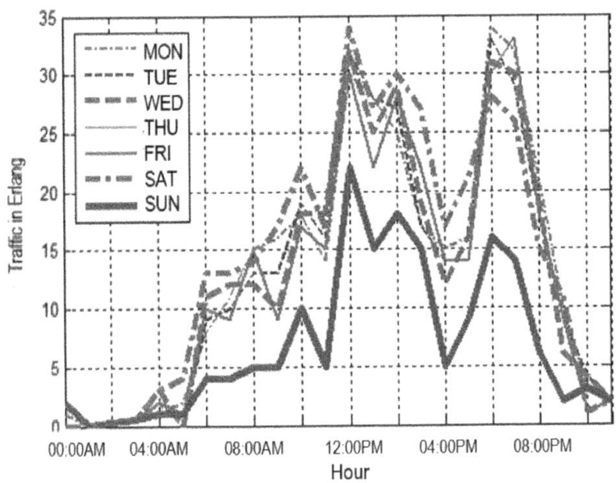

Figure 5.3. *Traffic carried by a GSM BTS in Antananarivo, Madagascar. For a color version of this figure, see www.iste.co.uk/rava/queues.zip*

All of the curves look similar. They represent the habits of people living or working in or near that environment. Traffic begins to rise from 5 am and reaches its first peak at around noon. This midday peak indicates the midday break from work when people leave their offices to find something to eat. Afterwards, traffic returns to its state from 10 am, when everyone was at work. The same phenomenon repeats itself at around 6 pm, making the second peak of the day. Afterwards, traffic falls to less than 5 Erl at around 11 pm, when most people, in bed, no longer use their phones.

In each case, a rather clear peak appears from around 6 to 8 pm. This peak is due, on the one hand, to habits, which dictate that callers wish to reach their interlocutors at these hours in order to avoid bothering them during working hours, and on the other hand, to the decrease in rates offered by the operator. Surprisingly, it seems that lower rates are not a very motivating factor for users: the operator will have chosen which hours to lower rates for accordingly!

The curve for Saturday is slightly higher during the day. It represents the weekly market when people meet others during the day as opposed to working hours. We can easily see that this traffic starts to decrease from 5 pm, when people start to go home.

The curve for Sunday represents a break. This creates a large dip in traffic on Sundays. The shape of the curve, however, remains the same.

5.3. Measuring and foreseeing traffic

5.3.1. *Traffic and service quality*

One of the objectives of teletraffic is to evaluate the risk of trivializing resources so they can be properly dimensioned for a better compromise between the cost of resources made available to users and the quality of the service they receive. Teletraffic is what provides the tools for planning a network, or the components of this network, for a profitable investment. Since carried traffic is the quantity that models the use of a network's resources, it is the only profitable component in it. We must thus carry as much traffic as possible on the network in question.

Deploying as many resources as possible can resolve the issue of carried traffic but, in terms of cost, this will inflate the investment. However, deploying insufficient resources risks causing some traffic to be lost, as explained in section 5.2.4.2. This impacts the quality of the service received by users, who have sought out a certain quality of service in exchange for their hard-earned money.

This quality of service includes a considerable number of parameters that determine whether a service will be considered good, acceptable or inacceptable. The principal parameter is, of course, the a priori probability that a request by a user will be accepted for a service, but this is far from the only one. To optimize a network, it is necessary to account for many different parameters. These service quality and carried traffic parameters are not only valid for telephony, but also, to different degrees, for any telecommunication network and for any type of service in telecommunications.

5.3.2. *Measuring traffic*

Teletraffic consists not only of estimating traffic but also of measuring it. To do this, we generally add all the durations h_i of all the M activities using resources during the period of observation T, then we relate the sum to this period T:

$$Y = \frac{1}{T}\sum_{i=1}^{M} h_i \qquad [5.4]$$

This follows from the definition that carried traffic is equal to the fraction of time during which resources are busy.

Equation [5.4] also allows us to interpret carried traffic as the average number of resources that are busy during the observation period T.

REMARK.– In the case where carried traffic Y represents traffic at the peak hour, the observation period T is equal to 1 hour.

The result Y of the measure of traffic at peak hours is quantifiable. We often compare it with nominal traffic Y_n corresponding to the theoretical maximal traffic that can be carried by a system (see the last two chapters of this volume):

– if the ratio Y/Y_n is less than 1, we say that the resources in the system are *under-used*. This system is *over-dimensioned*;

– if the ratio Y/Y_n is more than 1, we say that the resources in the system are *over-used*. This system is *under-dimensioned*;

– if the ratio Y/Y_n is close to 1, we say that the resources in the system are *correctly used*. This system is *appropriately dimensioned*.

5.3.3. *Markovian model of traffic*

Let us consider a set of N ($N \in \mathbb{N}^* \cup \{+\infty\}$) resources. At a given instant, k of these N resources are busy. These resources can be modeled by a Markov chain whose state is the number k of busy resources at a given instant. Each accepted request occupies a resource and the end of any activity frees a busy resource.

The frequency at which new requests appear corresponds to the rate of growth λ_k of the Markov chain in state k, and the frequency at which activity disappears corresponds to the death rate μ_k. We are thus dealing with a process of birth and death as described in section 2.3.

The calculation of the probabilities that these resources will be busy turns out to be the same as determining the stationary distribution of this birth and death process. Equation [2.36] provides the result for these activity probabilities. The probability of system blocking (all N resources are busy) is obtained from this stationary distribution of the birth and death process for state N. This equation does not give information about the nature of activity, but only its quantity. In certain cases, this limitation is not an issue since the problem of dimensioning that consisted of determining the number of necessary resources has already been resolved.

Besides this last point, other limits are also observed following the mutual independence hypothesis for request arrivals. This independence is no longer valid for repeat requests by the same user in the case of system blocking, nor is it for panic phenomena following unusual exterior events.

5.3.4. *Economy and traffic forecasting*

As mentioned in the introduction, the objective of the theory of teletraffic is to reduce the cost of a system. Forecasting is an important tool to reach this goal. In the case of telecommunication networks, this forecasting allows operators to calculate the potential cost of new network infrastructure or a new service for a given quality of service during the planning and conception stage, thus guaranteeing that costs will be reduced to the minimum.

The mathematical models established in this book are traffic forecasting tools. However, another important method used for this forecasting is simulation. It is the most common quantitative modeling technique used today. An important reason for this is, on the one hand, the complexity of mathematical traffic models, and, on the other hand, the calculating power of the computers doing the preferred simulation analysis for problems that are not easily resolved mathematically.

5.4. Exercises

1) Can you recite the definition of traffic?

2) In your opinion, on what basis should we estimate the traffic to be carried when planning a new line between two telephone exchanges?

3) Why is it necessary to choose the smallest possible interval of time for observation?

4) Can we process asynchronous transfer mode (ATM) traffic using Markov chains, and if so, how would we establish the mathematical model?

5) Can we process internet protocol (IP) traffic using Markov chains?

EXERCISE 5.1.– *On the telephone*

Let there be 10 users making 10-minute phone calls.

1) Calculate the total corresponding traffic by considering an observation time equal to 10 minutes, then an observation time of 30 minutes and finally an observation time of one hour.

2) Calculate the traffic per user for an observation time of one hour.

Let there be 100 users making one-minute phone calls.

3) Same questions.

4) What are the differences between these two cases?

EXERCISE 5.2.– *Road traffic*

Two cities A and B are 60 km apart.

1) Calculate the traffic generated by a car driving at an average speed of 80 km/h between the two cities A and B.

2) Calculate the traffic generated by 1,000 cars driving at an average speed of 80 km/h between the two cities A and B.

EXERCISE 5.3.– *Business telephone lines*

In a business, for a telephone line at the busy hour, the traffic can be broken down as follows:

– departure traffic: 0.035 Erl;

– arrival traffic: 0.045 Erl;

– internal traffic: 0.06 Erl.

1) Determine the total traffic per hour at the busy hour for one line and then for ten lines.

2) Determine the average duration of calls.

EXERCISE 5.4.– *Telephone booth*

On average, ten people arrive at a telephone booth every half hour. Each person makes, on average, a two-minute phone call in this booth.

1) Calculate the telephone booth's traffic.

During holidays, these arrivals are doubled.

2) Can the telephone booth accommodate all the traffic?

3) Determine the offered traffic.

4) Noting that the maximum traffic that the booth can accommodate is at most 1 Erl, determine the average number of arrivals that cannot be received by the booth.

EXERCISE 5.5.– *GSM*

Consider a very large airport covered by a telephone network. The maximum rate of arrival of travelers is 100,000 per hour. Let us suppose that 25% of travelers have cellular telephones that they turn on when they arrive and that, among them, 25% make a phone call of 60 seconds.

Let us suppose that the 60 seconds of communication are split into 5 seconds of SDCCH channel activity and 55 seconds of TCH channel activity. When the telephone is turned on, the SDCCH channel is used to update the location for 4 seconds.

1) Calculate the SDCCH channel traffic at the airport.

2) Calculate the TCH channel traffic at the airport.

6

Resource Requests and Activity

Waiting is also a kind of activity.

Cesare Pavese (1908–1950)

With this citation from the Italian author Cesare Pavese, we introduce this chapter devoted to resource requests and activity in any system. We consider continuous time and discrete time.

6.1. Infinite number of sources

6.1.1. *Distribution of requests in continuous time*

6.1.1.1. *Poisson process of requests*

In section 3.1.2, we showed that the process of appearance of rare events in a population of infinite size is a Poisson process.

A request can be considered a rare event since the appearance of a request at a given instant is low. The appearance of requests made by an infinite number of sources is therefore a Poisson process. We call them *Poissonian requests*.

This deduction necessitates the following a priori hypotheses:

– Hypothesis 1: in accordance with the amnesia of the Poisson process mentioned in Proposition 1.7, the probability of appearance of x requests in the interval between time t and $t + \Delta t$ depends only on the duration Δt of this interval of time, but not the starting time t. Moreover, the appearance of a new request is entirely fortuitous, without any relationship to what has happened in the past.

– Hypothesis 2: in accordance with the stationary principle of the Poisson process mentioned Proposition 1.6, there is no correlation between the number of requests received during different intervals of time.

– Hypothesis 3: in accordance with the simplicity of the Poisson process mentioned in Proposition 1.6, the probability of receiving more than one request is negligible compared to that of receiving one during a very short interval of time dt. The probability of receiving a request during this very short interval Δt is $p = c.dt$, where c is the average frequency of requests.

In everything that follows, dt designates a very short interval of time.

REMARK.– The average frequency of requests indicates the average number of requests per unit of time. If we use $E(\tau)$ to denote the expected value of the random variable τ of inter-arrivals in the process, we have $c = 1/E(\tau)$.

6.1.1.2. *Inter-arrivals*

In accordance with the inter-arrivals of the Poisson process of requests, we can also set out the following proposition.

PROPOSITION 6.1.– *The inter-arrivals τ of requests follow an exponential distribution with parameter c.*

PROOF.–

We can establish the complementary cumulative distribution function of τ in another manner from the hypotheses in section 6.1.1.1. For an inter-arrival τ greater than t, it is also greater than $t + dt$ with the probability:

$$\mathbb{P}(\tau > t + dt) = \mathbb{P}(\tau > t).(1 - c.dt)$$

where:

– $\mathbb{P}(\tau > t + dt)$ is the probability of an inter-arrival greater than $t + dt$, that is to say, the probability of no requests until $t + dt$;

– $\mathbb{P}(\tau > t)$ is the probability of no requests until t;

– $(1 - c.dt)$ is the probability of no requests between t and $t + dt$ using hypotheses 1 and 3.

Hypothesis 2 allows us to write $\mathbb{P}(\tau > t).(1 - c.dt)$ given the independence of these two events.

Thus:

$$\frac{\mathbb{P}(\tau > t + dt) - \mathbb{P}(\tau > t)}{dt} = -c.\mathbb{P}(\tau > t)$$

Now,

$$\frac{\mathbb{P}(\tau > t + dt) - \mathbb{P}(\tau > t)}{dt} = \frac{d}{dt}\mathbb{P}(\tau > t)$$

After integrating, we obtain:

$$\mathbb{P}(\tau > t) = Ke^{-ct}$$

Since $\mathbb{P}(\tau > t) = 1$, we find that $K = 1$. The complementary cumulative distribution function of the random variable τ is therefore:

$$\mathbb{P}(\tau > t) = e^{-ct}$$

The complementary cumulative distribution function of the random variable follows an exponential distribution with parameter c.■

Thus, the expected value of the inter-arrival equals $E(\tau) = 1/c$. Recall that c is the average frequency of requests.

6.1.1.3. *Number of requests*

In accordance with the counting measure of the Poisson process of requests, we can also set out the following proposition.

PROPOSITION 6.2.– *The number X of requests during an interval of time* Δt *follows a Poisson distribution with parameter* $c.\Delta t$.

PROOF.–

We can again establish this proposition starting with the hypotheses in section 6.1.1.1. We denote by $\mathbb{P}(X(\Delta t) = x)$ the probability of receiving x requests during an interval of time Δt.

The event "receiving x requests during an interval of time $\Delta t + dt$" is equivalent to:

– having x requests during Δt and none during dt, with a probability $\mathbb{P}(X(\Delta t) = x).(1 - c.dt)$;

– having $x-1$ requests during Δt and only one during dt, with a probability $\mathbb{P}(X(\Delta t)=x-1).(c.dt)$.

Therefore:

$$\mathbb{P}(X(\Delta t+dt)=x)=\mathbb{P}(X(\Delta t)=x).(1-c.dt)+\mathbb{P}(X(\Delta t)=x-1).(c.dt)$$

This is a differential equation for t and there are difference equations for x. Its solution is found by recurrence using $\mathbb{P}(X(\Delta t)=0)=e^{-c.\Delta t}$.

We leave it to the reader to expand it, but a Poisson distribution should be arrived at in the end:

$$\mathbb{P}(X(\Delta t)=x)=\frac{(c.\Delta t)^x}{x!}e^{-c.\Delta t} \blacksquare$$

Thus, the average number of requests during an interval of time Δt is equal to:

$$E\big(X(\Delta t)\big)=\sum_{x=0}^{\infty}x.\mathbb{P}(X(\Delta t)=x)=\sum_{x=0}^{\infty}\frac{(c.\Delta t)^x}{x!}e^{-c.\Delta t}$$
$$=c.\Delta t \qquad [6.1]$$

We find that the average number of requests is deducted from their average frequency c and the duration of the observation interval Δt.

6.1.2. *Distribution of requests in discrete time*

6.1.2.1. *Bernoulli process of requests*

This time, we consider discrete time. Numerical systems are governed by a clock. The appearance of any event only happens with the ticks of this clock. The instants considered are therefore multiple discrete instants of these ticks of a clock, not continuous.

Let there be a system governed by a clock with a period θ. The appearance of a request only happens over several ticks of the clock $k\theta$, $k \in \mathbb{N}^*$. We can adapt the hypotheses of section 6.1.1.1 as follows:

– Hypothesis 1: the probability of x requests appearing during the interval of time between t and $t+k\theta$ depends only on the number of ticks of the clock k in this interval of time, but not on the starting instant t. Moreover, the appearance of a new request is purely fortuitous, without any relationship to what has happened in the past.

– Hypothesis 2: there is no correlation between the number of the requests received during different intervals of time.

– Hypothesis 3: the probability of receiving more than one request is negligible compared to that of receiving one in the space of one tick of the clock. The probability of receiving one request at this instant equals p.

This type of process followed by requests was already set out in Exercise 1.11. It is called a Bernoulli process with parameter p. We can demonstrate that it is also a memoryless process like the Poisson process.

6.1.2.2. *Inter-arrivals*

PROPOSITION 6.3.– *The inter-arrivals* τ *of requests follow a geometric distribution with parameter* p.

PROOF.–

With this discrete time, the inter-arrival must also be a multiple of the ticks of the clock θ. The distribution of inter-arrivals τ is therefore defined by $\mathbb{P}(\tau = k\theta)$, which means no requests during $k - 1$ consecutive clock periods with a probability $(1 - p)$, and the request appears at the k-th period with a probability p.

Indeed, according to hypothesis 3, at a given instant (which is a priori a multiple of the tick of a clock), the probability of a request appearing is p.

According to hypothesis 2, we can write the product $\mathbb{P}(\tau = k\theta) = (1 - p)^{k-1}.p$, which is none other than the probability distribution of a random variable following a geometric distribution with parameter p.■

Thus, the average inter-arrival is equal to:

$$E(\tau) = \sum_{k=0}^{\infty} \tau.\mathbb{P}(\tau = k\theta) = \sum_{k=0}^{\infty} k\theta.(1-p)^{k-1}.p = \frac{\theta}{p} \qquad [6.2]$$

The average frequency of requests is:

$$c = \frac{1}{E(\tau)} = \frac{p}{\theta} \qquad [6.3]$$

6.1.2.3. *Number of requests*

PROPOSITION 6.4.– *The number X of requests during an interval of time $k\theta$ follows a binomial distribution with parameter (k,p) or $(k,c\theta)$.*

PROOF.–

We denote by $\mathbb{P}(X(k\theta)=x)$ the probability of receiving x requests during an interval of time $k\theta$.

The event "receiving x requests during k clock periods" is equivalent to:

– having x requests in x clock periods, each with a probability p;

– not having any requests in $k-x$ clock periods, with a probability $1-p$;

– x combinations from k for the x clock periods where the requests appear.

Therefore:

$$\mathbb{P}(X(k\theta)=x)=\binom{k}{x}p^x(1-p)^{k-x}$$

We find the probability distribution for a random variable following the binomial distribution with parameter k and $p=c\theta$.■

Thus, the average number of requests during an interval of time $k\theta$ is equal to:

$$E\big(X(k\theta)\big)=\sum_{x=0}^{\infty}x.\mathbb{P}(X(k\theta)=x)=kp=c.k\theta \qquad [6.4]$$

Once more, we find that the average number of requests is deducted from their average frequency c and from the duration of the observation interval $k\theta$.

REMARK.– For a clock tick tending towards zero, the system in discrete time tends towards a system in continuous time; the Bernoulli process tends towards the Poisson process. Indeed, the geometric distribution of inter-arrivals tends towards an exponential distribution, and the binomial distribution of the number of requests tends towards the Poisson distribution.

6.1.3. *Duration of activity distributions*

6.1.3.1. *Duration of i.i.d. services*

The process of the duration of services is generally complex. This process can depend on numerous factors that are external to the system: the history of the process, the appearance time of the process, the type of service that is the object of the process, the history of previously processed requests, etc. We can even extend it to whether there are priority requests that can preempt resources being used by another, therefore shortening the duration of the service in progress.

For mathematical analyses, we are compelled to posit certain hypotheses capable of leading to interpretable results. This allows us to simplify the problem while avoiding the complexities that arise, but it is not adequate in a more general case. For that, we can adopt the following hypotheses:

– Hypothesis 1: the different durations of activity follow the same distribution.

– Hypothesis 2: the different durations of activity are independent of each other.

In other words, we accept that the different durations of activity are independent and identically distributed.

These hypotheses reflect the observed reality in the case of telephone traffic and other cases of mono-service traffic. For the case of a multiservice such as the Internet, hypothesis 1 is already difficult to accept.

6.1.3.2. *Exponential duration of service*

Let D be the random variable for duration of service. We will consider an interval of length of d that we divide into k sub-intervals of duration d/k. The number k of sub-intervals can be chosen so that:

– the probability that activity will end within a sub-interval is proportional to the duration of this sub-interval. We will denote this probability $\mu.d/k$, where μ is a coefficient of proportionality;

– the probability that activity will end within a sub-interval is independent of this sub-interval.

Thus, the event "$D > t$", which means that activity beginning at instant $t = 0$ has not yet ended at instant t (duration $d = t$), is equivalent to the event "activity does not end within any of the k sub-intervals". Let us recall that the probability that activity will not end within a sub-interval is equal to $1 - \mu.t/k$. We thus have:

$$\mathbb{P}(D > t) = \left(1 - \frac{\mu t}{k}\right)^k \qquad [6.5]$$

We must divide the duration $d = t$ into an infinite number of sub-intervals in order to have an activity endpoint that can appear at any instant. By having k tend towards infinity, we obtain:

$$\mathbb{P}(D > t) = \lim_{k \to +\infty} \left(1 - \frac{\mu t}{k}\right)^k = e^{-\mu t} \qquad [6.6]$$

We thus find the complementary cumulative distribution function of an exponential distribution with parameter μ, or with an average $E(D) = 1/\mu$. We already discussed in section 1.1.1 how the exponential distribution allows us to model lifespans, but only without aging due to its memoryless property.

This observation, which comes down to the behavior of users, may appear somewhat surprising. Indeed, we can deduce that the probability of a telephone conversation ending does not depend on the lifespan of the telephone conversation, although, intuitively, we might instead expect this probability to rise with time. In other words, it appears that this duration has some level of aging. These intuitive observations are not entirely baseless: it is simply the case that an exponential distribution is sufficiently realistic for our needs, with the added benefit of simplicity.

Statistical observations of real cases indicate that resource activity that is directly tied to human behavior has a duration d whose statistical distribution presents a clearly weaker maximum than the average value $E(D)$, and that this distribution slowly decreases for increasing durations. It is the exponential distribution that allows us to perfectly model this behavior.

6.1.3.3. *Constant duration of service*

Another way of modeling durations of service time can also be considered, specifically, a deterministic duration. Many examples can be found, such as access to a centralized computer, the time a packet spends in transit, etc. It is not, however, as realistic as has been thought, but it allows us to simplify an almost-constant model of duration.

Hypothesis 2 set out in section 6.1.3.1 is no longer valid in this case due to the complete correlation of activity duration in two disconnected sub-intervals.

$$D = E(D) = constant \qquad [6.7]$$

6.1.4. *Distribution of busy sources*

We use S to denote the random variable equal to the number of sources that are simultaneously busy in an infinite population of sources offering an offered traffic A.

PROPOSITION 6.5.– *The number S of simultaneously busy sources offering an offered traffic A follows a Poisson distribution with parameter A.*

PROOF.–

We use N to denote the number of sources, and then we have it tend towards infinity. To generate traffic N, $S = s$ sources must be simultaneously busy, with the probability of just one source being busy being equal to A/N (recall the definition of offered traffic). A source can be free (probability $1 - A/N$) or busy (probability A/N).

Moreover, the states of sources, and whether they are busy or not, are independent of each other, and there is the possibility of s sources from N being combined to select these s simultaneously busy sources.

Thus, the probability of finding s simultaneously busy sources is equal to:

$$\mathbb{P}(S = s) = \binom{N}{s}\left(\frac{A}{N}\right)^{s}\left(1 - \frac{A}{N}\right)^{N-s}$$

For N tending towards infinity and a fixed A, the measure A/N tends towards 0 and this probability tends towards a Poisson distribution, which is the limit of the binomial distribution:

$$\mathbb{P}(S = s) = \lim_{N\to+\infty}\binom{N}{s}\left(\frac{A}{N}\right)^{s}\left(1 - \frac{A}{N}\right)^{N-s} = \frac{A^{s}}{s!}e^{-A}$$

We accordingly arrive at a Poisson distribution with parameter A.■

Thus, the average number of simultaneously busy sources is equal to $E(S) = A$. The traffic offered is equal to the average number of simultaneously busy sources needed to provide it. We recall that offered traffic A is equal to $A = c_A h = \lambda/\mu$.

6.2. Finite number of sources

The number of sources is not always infinite. There are cases where we must consider finite numbers of sources, such as in the case of dimensioning an internal network at a company. In this section, we consider a finite number of sources requesting resources from the system, offering an offered traffic A.

In this case, the hypotheses of the Poissonian requests are no longer valid. Indeed, the arrival process of requests depends on the past requests, and in particular on the number of those accepted or not. The probability of a new request naturally falls when the number of busy sources rises, and this is due to the fact that the number of sources that will be able to make a request diminishes.

6.2.1. *Modeling with birth and death processes*

Let there be a system with a finite number N of sources. The number of resources in this system is denoted n. A source will use only one resource for its service once its request is accepted.

We can model such a system with a process of birth and death whose states are defined by the random variable S designating the number of busy sources. The states that the random variable S can take are integers between 0 and N.

A new request appears with a probability depending on the number of sources that are already busy. Denoting the rate of appearance of requests as λ_k when k sources are busy, the probability of a new request appearing in an infinitely small interval of time ε is equal to $\lambda_k \varepsilon + o(\varepsilon)$ (see the birth and death process as described in section 2.3).

If we use c to denote the average frequency of a request, the rate of appearance of a request, which is a function of the number of free sources $N - k$, is equal to:

$$\lambda_k = c.(N - k) \qquad [6.8]$$

An accepted request will use a resource for the duration of a service assumed to be i.i.d. random variables from the exponential distribution with an average of h, given the independence of the services times of different sources. Using μ_k to denote the rate at which service endpoints appear when k resources are busy (with the k sources), the probability an activity will end in an infinitely small interval of time ε will be equal to $\mu_k \varepsilon + o(\varepsilon)$. We have:

$$\mu_k = \frac{k}{h} \qquad [6.9]$$

Indeed, the probability that only one resource will cease activity is equal to $\frac{1}{h}\varepsilon + o(\varepsilon)$, and we have k independently busy resources.

Figure 6.1 represents the transitions between states in this birth and death process described in Equations [6.8] and [6.9].

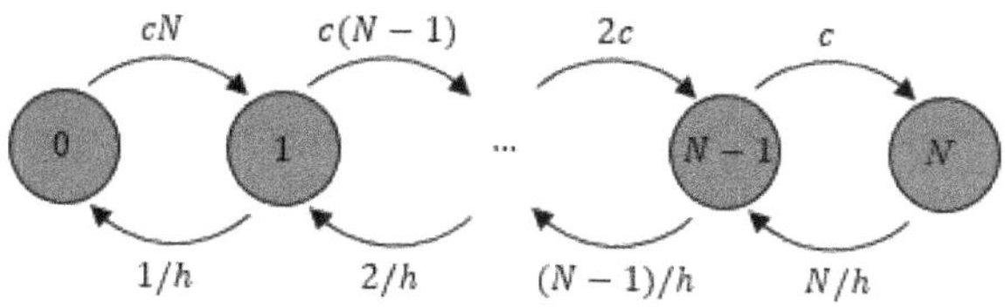

Figure 6.1. *Birth and death process of requests*

6.2.2. *Distribution of requests*

PROPOSITION 6.6.– *The number of simultaneously busy sources S offering an offered traffic A follows a binomial distribution with parameter N and* A/N.

PROOF.–

By calculating the stationary distribution of the birth and death process of the number of busy sources, we have, according to Equation [2.36]:

$$\begin{aligned}\pi_k &= \frac{\lambda_0\lambda_1 \dots \lambda_{k-1}}{\mu_1\mu_2 \dots \mu_k}\pi_0, \qquad k = 1,2,\dots,N \\ &= \frac{c.N.c.(N-1)\dots c.(N-k+1)}{\frac{1}{h}\cdot\frac{2}{h}\cdots\frac{k}{h}}\pi_0 \\ &= \frac{(ch)^k.N(N-1)\dots(N-k+1)}{k!}\pi_0 = \frac{(ch)^k N!}{k!(N-k)!}\pi_0 \\ &= \binom{N}{k}(ch)^k\pi_0\end{aligned}$$

The definition of the stationary distribution $\sum_{i=0}^{N}\pi_i = 1$ gives us:

$$\frac{1}{\pi_0} = \sum_{i=0}^{N}\binom{N}{i}(ch)^i = (1+ch)^N$$

Thus, we obtain:

$$\pi_k = \binom{N}{k}\frac{(ch)^k}{(1+ch)^N} = \binom{N}{k}\left(\frac{ch}{1+ch}\right)^k\left(\frac{1}{1+ch}\right)^{N-k}$$

$$= \binom{N}{k} \left(\frac{ch}{1+ch}\right)^k \left(1 - \frac{ch}{1+ch}\right)^{N-k}$$

Note that ch is the average offered traffic per source since c is the average frequency of requests and h is the average duration of activity. Now, the set of N sources offers a total traffic A; therefore, the probability that one source will be busy is A/N and that it will be free is $1 - A/N$. We must therefore have:

$$\frac{A}{N} = ch.\left(1 - \frac{A}{N}\right)$$

$$ch = \frac{A}{N - A}$$

and:

$$\frac{ch}{1+ch} = \frac{\frac{A}{N-A}}{1+\frac{A}{N-A}} = \frac{A}{N}$$

In that case, we can write the stationary distribution as:

$$\pi_k = \binom{N}{k} \left(\frac{A}{N}\right)^k \left(1 - \frac{A}{N}\right)^{N-k}$$

We arrive at the binomial distribution with parameter N and A/N.■

Thus, we once more find that offered traffic represents the average number of busy sources. Indeed, we have $E(\pi_k) = N.\frac{A}{N} = A$.

6.3. Traffic peaks and randomness

6.3.1. *Traffic peaks*

DEFINITION 6.1.– *What we call the peakedness factor of traffic Z is the relationship between the variance and the average value of traffic:*

$$Z = \frac{\sigma^2(X)}{E(X)} \qquad [6.10]$$

where X represents either the traffic A offered by sources or the traffic Y carried by resources.

It represents the dispersion of traffic relative to its average value.

The traffic is called *smooth* (or suppressed) when its peakedness factor is smaller than 1. It is called *peak* (or induced) when its peakedness factor is greater than 1.

REMARK.– The peakedness factor can also be calculated from the stationary distribution of the number of busy sources or resources. It is defined by the relationship between its variance and its average value.

6.3.2. *Pure chance traffic*

Let us consider a system serving its customers for durations of service with an exponential distribution with an average of $1/\mu$. The traffic process is a birth and death process, which is a particular type of Markov chain. This traffic process is called *insensitive*, that is, only the average duration of service impacts the stationary distribution of the system, but not the type of this distribution.

The process of arrival of customers is considered a Poisson process with intensity λ. The type of traffic obtained is called *random traffic* or *pure chance traffic type-one*. In this case, the peakedness factor Z is equal to 1.

In the case of a finite number of sources N, a free, individual source can create a request with an arrival intensity of λ. Once busy, it can no longer create a request, so its arrival intensity is 0. The arrival process therefore depends on the state of the system, and, using k to denote the number of busy sources, we get at $(N-k)\lambda$ the intensity arrival of the process. The type of traffic obtained is random traffic or *type-two pure chance traffic*. In this case, the peakedness factor Z is smaller than 1.

6.4. Recapitulation

Table 6.1 summarizes the arrival process of requests in continuous time as well as that governed by a clock with a period of θ. Note that all the characteristics of the process in discrete time tend towards those of the process in continuous time when the period of the clock tends towards zero.

Table 6.2 summarizes the distributions of source activity in terms of duration and quantity.

	Continuous time	**Discrete time**
Process of arrival of requests	Poisson	Bernoulli
Instants of requests	At any instant	At most one per clock period θ
Distribution of inter-arrivals τ	Exponential with parameter c	Geometric with parameter p
Average inter-arrival	$E(\tau) = \frac{1}{c}$	$E(\tau) = \frac{\theta}{p} = \frac{1}{c}$
Distribution of the number of requests X in the interval of time Δt	Poisson with parameter $c.\Delta t$	Binomial with parameter k and $c.\theta$, where $\Delta t = k\theta$
Average number of requests during Δt	$E(X) = c.\Delta t$	$E(X) = ck\theta = c.\Delta t$

Table 6.1. *Continuous and discrete arrival processes*

	Infinite sources	**Finite sources**
Distribution of activity duration T	Exponential with parameter μ	Exponential with parameter μ
Average duration of activity	$E(T) = 1/\mu$	$E(T) = 1/\mu$
Distribution of busy sources S offering an offered traffic A	Poisson with parameter A	Binomial with parameter N and A/N
Average number of busy sources offering an offered traffic A	$E(S) = A$	$E(S) = N.\frac{A}{N} = A$

Table 6.2. *Distribution of source activity*

6.5. Exercises

1) What does the rate of requests measure?

2) What does the Poisson distribution measure? Under what hypotheses is this distribution valid?

3) What are the differences between the Poisson and the Bernoulli distributions?

4) An SMB purchases a domestic telephone exchange to connect its six partners to the public telephone network. Can the traffic of the partners can be estimated by the Poisson formula? Why?

EXERCISE 6.1.– *Infinite source traffic*

Let there be traffic of 5 Erl offered by sources of infinite size.

1) Calculate the probability that this traffic will be generated by just one source.

2) Calculate the probability that this traffic will be generated by five sources.

3) Five sources generate this traffic for an average of one hour of activity. What is the probability that the activity requested by a source is greater than or equal to 30 minutes?

EXERCISE 6.2.– *Busy sources*

A request arrives from a source every 10 minutes on average. Let us suppose an infinite number of sources:

1) Calculate the average number of requests in one hour.

2) Calculate the probability that six requests will arrive in one hour.

3) Each source has a probability equal to half of the requesting activity of more than 30 minutes. Supposing an exponential distribution of activity duration, calculate its average duration.

4) Calculate the traffic offered by these sources.

5) Deduce the average number of busy sources.

EXERCISE 6.3.– *Printer*

A small printing company has five employees. They use shared printers in their daily work.

1) In one out of three cases, one employee uses the printers for more than one minute. Calculate the average duration of use considering an exponential distribution of this activity.

2) For an offered traffic of 3 Erl of printer activity, calculate the probability that this traffic is offered by all of the employees simultaneously.

3) What is the average number of busy employees offering this traffic of 3 Erl?

4) Calculate the average traffic per busy employee, then the average traffic per unoccupied employee.

7

The Teletraffic of Loss Systems

To save one's credit, one must hide one's losses.

Jean de la Fontaine (1621–1695)

In Chapter 6, we examined distributions of requests and durations of activity. However, the principal objective of teletraffic is the evaluation of the number of resources needed to provide customers with a satisfactory quality of service. This chapter has the objective of evaluating this quantity using the distributions of the numbers of busy resources for different scenarios involving loss systems.

For type-one pure chance traffic (PCT-1), two scenarios can exist:

– infinite number of resources n: Poisson distribution;

– finite number of resources n: truncated Poisson distribution or Erlang distribution.

For type-two pure chance traffic (PCT-2), two scenarios can also exist:

– number of resources $n \geq N$ sources: binomial distribution;

– number of resources $n < N$ sources: truncated binomial or Engset[1] distribution.

1 Thorir Olaus Engset (1865–1943), Norwegian mathematician and engineer.

7.1. Loss systems

7.1.1. *Definitions*

DEFINITION 7.1.– *What we call a perfect system is a system in which free resources are always accessible.*

Given Definition 7.1, the system can allocate one of its unoccupied resources. A perfect system is also called a system with perfect access, or with *total accessibility*.

DEFINITION 7.2.– *What we call an imperfect system is a system in which an unoccupied resource is not always necessarily accessible.*

In the case of this imperfect system, there can be internal blockages, of which we can distinguish two types:

– limited but constant accessibility: resource accessibility does not depend on traffic offered to the system but itself possesses resources that are inaccessible from the outside;

– variable accessibility: resource accessibility depends of the traffic offered to the system. This accessibility is perfect when the offered traffic is equal to zero, and it diminishes when the traffic increases.

These two systems are called *loss systems* since they reject the traffic they cannot carry. This rejected traffic is irrevocably lost.

7.1.2. *Blocking and loss*

7.1.2.1. *Blocking probability*

DEFINITION 7.3.– *What we call blocking is the state of a perfect system (or its resources) in which all its resources are busy at the same time.*

In the case of blocking, potential new requests can no longer be processed by the system since it no longer has resources to server the requesting customer.

The blocking probability is denoted B. Using X to denote the number of busy resources among the initially available n resources in a system, it is equal to $B = \mathbb{P}(X = n)$.

According to the property of ergodicity, we can also say that it is the percentage of time during which the system is observed to be in a state of blocking with respect to a period of observation.

The blocking probability is also called the *time congestion*.

7.1.2.2. *Loss probability*

DEFINITION 7.4.– *The loss probability is the probability that a request might not be processed and will be irrevocably lost.*

The loss probability is denoted E.

From Definition 7.4, we can say that the loss probability is relative to requests. In other words, it is the relationship between the number of refused requests and the total number of requests made during a period of observation. If we denote the number of refused requests per unit of time c_R, we can write that the loss probability is equal to the ratio of refused traffic to offered traffic:

$$E = \frac{c_R}{c} = \frac{c_R h}{ch} = \frac{A_p}{A} \quad [7.1]$$

where A_p designates refused traffic: $A_p = c_R.h = A - Y$.

Consequently, we have:

$$E = 1 - \frac{Y}{A} \quad [7.2]$$

The traffic that is actually carried by the resources of a system can thus be written in terms of the traffic and the loss probability using the following expression:

$$Y = A.(1 - E) \quad [7.3]$$

The difference $A_p = A - Y = AE$ gives rejected traffic or lost traffic.

The loss probability is also called the *call congestion*.

7.1.2.3. *Comparison*

In a state of congestion, all of the resources in a system are busy at the same time. At that moment, there has still not been any loss; it only appears at the instant when a new request arrives while in this state. Consequently, the loss probability depends on the simultaneous arrival of this new request with a state of resource congestion.

Using NS to denote the event "a request arrives", we can write:

$$E = \mathbb{P}(X = n|NS) \tag{7.4}$$

We must therefore have $E \leq B$.

PROPOSITION 7.1.– *For the case of type-one pure chance traffic, the blocking probability is equal to the loss probability.*

PROOF.–

From equation [7.4], we have:

$$E = \mathbb{P}(X = n|NS) = \frac{\mathbb{P}(X = n \cap NS)}{\mathbb{P}(NS)}$$

Now, for type-one pure chance traffic, requests are Poissonian, that is, their process of arrival is memoryless: the probability of a new request does not depend on the requests already made in the past, even those which caused n resources to be busy. We must therefore have $\mathbb{P}(X = n \cap NS) = \mathbb{P}(X = n).\mathbb{P}(NS)$.

Thus,

$$E = \frac{\mathbb{P}(X = n \cap NS)}{\mathbb{P}(NS)} = \frac{\mathbb{P}(X = n).\mathbb{P}(NS)}{\mathbb{P}(NS)} = \mathbb{P}(X = n) = B$$

The loss probability is therefore equal to the blocking probability in this case.■

REMARK.– The following two hypotheses are called "lost calls cleared". They are the hypotheses proposed by Erlang in his model:

– requests appearing in a state of congestion are refused;

– they disappear instantly and do not lead to a new attempt.

Other hypotheses proposed by Molina[2], called "lost calls held", can also be considered in the calculation of loss systems.

7.2. The Erlang model

The Erlang model consists of a system with n identical resources working in parallel, with type-one pure chance traffic. The number of sources offering traffic to

2 Edward Charles Dixon Molina (1877–1964), American engineer.

the system is infinite. An arriving request is accepted for a service if a resource in the system is free.

Depending on the number of resources n, two cases can exist:

– an infinite number of resources $n = \infty$: the distribution of the number of busy resources follows a Poisson distribution;

– a finite number of resources $n < \infty$: the distribution of the number of busy resources follows a truncated Poisson distribution, also called an *Erlang distribution*.

7.2.1. *Infinite number of resources*

The traffic of the Erlang model is type-one pure chance traffic, and requests arrive according to a Poisson process of a fixed intensity $\lambda = c$. Since we have an infinite number of resources, any request arriving will therefore be accepted. The duration of activity follows an exponential distribution whose mean is $1/\mu = h$.

From Definition 5.14 of offered traffic, we can say that it is equal to the traffic carried by the system when the number of resources is equal to infinity: $A = ch$.

PROPOSITION 7.2.– *The number of resources R that are simultaneously busy with offered traffic A follows a Poisson distribution with parameter A:*

$$\mathbb{P}(R = k) = \frac{A^k}{k!}e^{-A} \quad [7.5]$$

PROOF.–

A system of n resources can be modeled by a Markov chain whose system state is the number of busy resources: $R \in \mathbb{N}$. The transition from state $R = k$ to state $R = k + 1$ is equivalent to the arrival of a request, and the transition from state $R = k$ to state $R = k - 1$ is equivalent to a resource ceasing activity.

We thus find ourselves with a birth and death process whose birth intensity is equal to λ whatever the state k of the system may be. The death rate of the system is equal to $\mu_k = k\mu$ if its state is equal to k.

Referring to equation [2.36] with $\lambda_k = \lambda = c$ and $\mu_k = k\mu = k/h$, we obtain:

$$\pi_k = \frac{\lambda_0\lambda_1 \dots \lambda_{k-1}}{\mu_1\mu_2 \dots \mu_k}\pi_0 = \frac{c^k}{k!\left(\frac{1}{h}\right)^k}\pi_0 = \frac{A^k}{k!}\pi_0, \qquad k = 1, 2, \dots$$

With $\sum_{i=0}^{+\infty} \pi_i$, we can derive:

$$\frac{1}{\pi_0} = \sum_{i=0}^{+\infty} \frac{A^i}{i!} = e^A$$

and:

$$\pi_k = \frac{A^k}{k!}e^{-A}$$

We find that the stationary distribution of the number of busy resources follows a Poisson distribution whose parameter is equal to the offered traffic A.■

The average number of busy resources is given by $E(R) = A$. For an infinite number of resources, the traffic carried by the system is equal to the offered traffic:

$$Y = A \qquad [7.6]$$

No traffic is lost in this case. The blocking probability is equal to the loss probability, and both are equal to zero:

$$B = E = 0 \qquad [7.7]$$

Since the variance of the Poisson distribution with parameter A is equal to A, the peakedness factor of the traffic is equal to $Z = 1$.

This model with an infinite number of resources is not as useful in terms of dimensioning. We will limit the number of resources in order to evaluate how the Erlang model performs.

7.2.2. Finite number of resources

Let us consider the same model as the one for type-one pure chance traffic, but this time with a finite number of resources n.

PROPOSITION 7.3.– *The number of resources R that are simultaneously busy with offered traffic A follows a truncated Poisson distribution with parameters A and n. This truncated Poisson distribution is also called an Erlang distribution.*

$$\mathbb{P}(R = k) = \frac{\frac{A^k}{k!}}{\sum_{i=0}^{n} \frac{A^i}{i!}} = \frac{\frac{A^k}{k!}}{1 + A + \frac{A^2}{2!} + \cdots + \frac{A^n}{n!}} \quad [7.8]$$

PROOF.–

We use the proof from Proposition 7.2 but with a finite number of resources n instead of an infinite number, that is, $\sum_{i=0}^{n} \pi_i = 1$, with $1/\pi_0 = \sum_{i=0}^{n} A^i/i!$ and $\pi_k = \frac{A^k}{k!}/\sum_{i=0}^{n} \frac{A^i}{i!}$. ■

We call this distribution "a truncated Poisson distribution" because we have truncated it to $i = n$, the sum of the denominator, which should tend towards e^A. In order to avoid confusion, the term "Erlang distribution" is restricted to sums of independent variables with exponential distributions, as described in section 1.1.2.3.

The blocking probability is obtained when $k = n$:

$$B = B_{1,n}(A) = \frac{\frac{A^n}{n!}}{1 + A + \frac{A^2}{2!} + \cdots + \frac{A^n}{n!}} \quad [7.9]$$

Equation [7.9] is called the *Erlang-B formula* or the *first Erlang formula*.

The loss probability is equal to the blocking probability because we have Poissonian requests:

$$E = B = B_{1,n}(A) \quad [7.10]$$

Carried traffic is equal to:

$$Y = A.\left(1 - B_{1,n}(A)\right) \quad [7.11]$$

Lost traffic is equal to:

$$A_p = A.E_{1,n}(A) \quad [7.12]$$

The improvement factor indicates the additional carried traffic when the number of resources is increased from n to $n+1$. Denoting carried traffic Y_n if we have n resources, we get:

$$F_{1,n}(A) = Y_{n+1} - Y_n = A.(B_{1,n}(A) - B_{1,n+1}(A)) \qquad [7.13]$$

The variance of the distribution of the number of busy resources is equal to:

$$\sigma^2(R) = \sum_{k=0}^{n} k^2\pi_k - Y^2 = Y - (n-Y)(A-Y) \qquad [7.14]$$

Since $Y \leq n$ and $Y < A$, $\sigma^2(R)$ is less than Y, which is the average of R, and the peakedness factor of the traffic is less than 1.

According to the comparison in Figure 7.1, we can see that the Erlang and Poisson distributions are similar. But while the Poisson distribution extends towards infinity, the Erlang distribution is truncated at $x = n$, and the values are majorized so as to give a sum of probabilities equal to 1.

The similarity is very visible for a small load, while the Erlang distribution is always greater than the Poisson one.

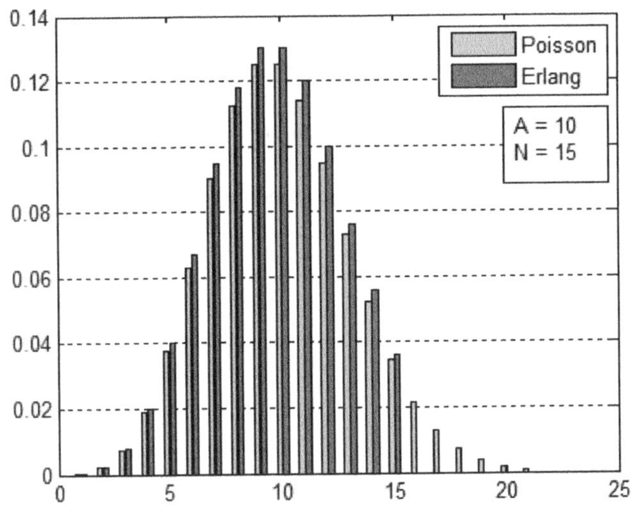

Figure 7.1. *Comparison of Erlang and Poisson distributions. For a color version of this figure, see www.iste.co.uk/rava/queues.zip*

7.2.3. *Erlang-B formula*

As already explained, equation [7.9] is called the Erlang-B formula. The blocking probability B is equal to the loss probability E.

$E = E_{1,n}(A)$ depends on A and n. For dimensioning, we should solve for n with respect to E (or B) and A, which is not possible analytically. The value of n can be found using a table called the *Erlang-B table* (see Appendix 2) or by using an abacus like the one in Figure 7.2. An estimate from approximation formulas can also be useful in certain cases.

We can also demonstrate the recursive formula:

$$B_{1,n}(A) = \frac{A.B_{1,n-1}(A)}{x + A.B_{1,n-1}(A)}, \quad B_{1,0}(A) = 1 \qquad [7.15]$$

This formula is rather useful when we wish to solve for $B_{1,n}(A)$ for large values of n given that $n!$ and A^n grow very quickly.

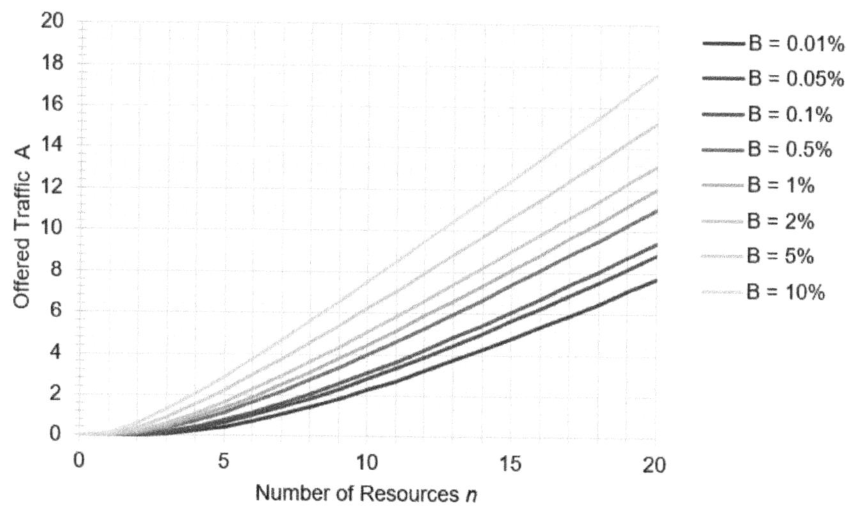

Figure 7.2. *Erlang-B chart. For a color version of this figure, see www.iste.co.uk/rava/queues.zip*

This recurrence equation can also be written in a linear form:

$$\frac{1}{B_{1,n+1}(A)} = 1 + \frac{n+1}{A}.\frac{1}{B_{1,n}(A)}, \quad B_{1,0}(A) = 1 \qquad [7.16]$$

We can simplify [7.16]:

$$\frac{1}{B_{1,n}(A)} = 1 + \frac{n}{A} + \frac{n(n-1)}{A^2} + \cdots + \frac{n!}{A^n} \quad [7.17]$$

The term $k(k-1)\ldots/A^k$ approaches zero when the value of k augments, so for an approximation, we can stop at a certain value for k.

7.2.4. *Dimensioning principles*

Estimating the quantity of resources in a loss system is generally undertaken with the purpose of obtaining the smallest possible loss probability. The ideal is assuredly to have this probability B be zero, but this implies an infinite number of resources, as shown in equations [7.7] and [7.10]. We must therefore reasonably impose an admissible value on this probability so we can obtain a dimensioning estimate.

This dimensioning consists of determining the number of resources n that allow for a maximum loss probability $E = E_n(A)$ for an offered traffic A. However, it is very difficult to find the exact value of n from equation [7.10]. Only tables or an abacus will allow us to determine it.

Table 7.1 shows the different traffic that can be offered to a system with n resources in order to obtain a loss probability of 1%. The average traffic offered per resource is indicated by a.

n	1	2	5	10	20	50	100
$A\ (B = 1\%)$	0.010	0.153	1.361	4.461	12.03	37.90	84.06
a	0.010	0.076	0.269	0.442	0.596	0.750	0.832
$F_{1,n}(A)$	0.000	0.001	0.011	0.027	0.052	0.099	0.147
$A_1 = 1,2.A$	0.012	0.183	1.633	5.353	14.44	45.48	100.9
$B\ (\%)$	1.198	1.396	1.903	2.575	3.640	5.848	8.077
a	0.012	0.090	0.320	0.522	0.696	0.856	0.927
$F_{1,n}(A_1)$	0.000	0.002	0.023	0.072	0.173	0.405	0.617

Table 7.1. *Offered traffic, loss probabilities and improvement functions*

Using Table 7.1, we can determine the number of necessary resources and evaluate the impact of variations in traffic on the loss probability.

Table 7.1 is only an example. A more complete table, the Erlang-B table, is provided at the end of this book.

EXAMPLE.– To carry an offered traffic of 12 Erl, we need 20 resources for a loss probability of 1%. With a variation of 10% in the traffic, the loss probability can reach up to 3.6%.

REMARK.– Table 7.1 shows the importance of the quantity of resources in a system (we sometimes call this the beam size). The analysis of traffic per resource allows us to conclude that a system with numerous resources is more efficient than numerous systems with less resources, even though the total number of resources is identical. In other words, more traffic can be carried if we use a bigger beam. Indeed:

– setting up 100 systems with one resource each allows us to carry an offered traffic of 1 Erl (100×0.010) with 1% loss, and the efficiency is $a = 0.01$;

– setting up 50 systems with two resources each allows us to carry an offered traffic of 7.63 Erl (50×0.153) with 1% loss, and the efficiency is $a = 0.269$;

– setting up 1 one system with 100 resources allows us to carry an offered traffic of 84 Erl (1×84.064) with 1% loss, and the efficiency is $a = 0.83$.

7.3. Engset model

The Engset model consists of a system of n identical resources working in parallel, with type-two pure chance traffic. The number of sources offering traffic to the system is in this case denoted N. An arriving request is accepted for service if a resource in the system is free.

Depending on the number of n resources, two cases can occur:

– for a sufficient number of resources $n \geq N$: the distribution of the number of busy resources follows a binomial distribution;

– for an insufficient number of resources $n < N$: the distribution of the number of busy resources follows a truncated binomial distribution, also called the *Engset distribution*.

7.3.1. *Sufficient number of resources*

Let us reconsider the model from section 6.2 for $n \geq N$. All arriving requests are accepted and result in resource activity. The number of simultaneously busy resources can take a value $0, 1, \ldots, n$.

Resource activity can be modeled by a birth and death process identical to the one in section 6.2.1. The rate of growth of the process is equivalent to that of the arrival of a request (immediately accepted), which is equal to $\lambda_k = (N-k)\lambda = (N-k)c$, where k designates the number of busy resources (equivalent to the number of busy sources). The death rate of the process corresponds to the cessation of activity among the current k resources in activity: $\mu_k = k\mu = k/h$.

PROPOSITION 7.4.– *The number of resources R that are simultaneously busy with offered traffic A follows a binomial distribution with parameter N and A/N:*

$$\mathbb{P}(R=k) = \binom{N}{k}\left(\frac{A}{N}\right)^k\left(1-\frac{A}{N}\right)^{N-k} \qquad [7.18]$$

PROOF.–

We prove equation [7.18] as we did Proposition 6.6 in section 6.2.2.■

The average number of busy resources is given by $E(R) = N.\frac{A}{N} = A$. For a sufficient number of resources $n \geq N$, the traffic carried by the system is equal to the offered traffic:

$$Y = A \qquad [7.19]$$

No traffic is lost in this case. The blocking probability is equal to the loss probability, both of which are equal to zero, for $n > N$:

$$B = E = 0 \qquad [7.20]$$

When $n = N$, the loss probability is still $B = 0$, but the blocking probability is equal to:

$$B = \left(\frac{A}{N}\right)^n \qquad [7.21]$$

Since the variance of the binomial distribution with parameters N and A/N is equal to $N.A/N.(1-A/N)$, the peakedness factor of the traffic is then worth $Z = 1 - A/N$, less than 1.

This model with a sufficient number of resources is not yet useful in terms of dimensioning. We can reduce the number of resources with respect to the number of sources to observe how Engset model performs.

7.3.2. *Insufficient number of resources*

This time, let us consider an insufficient number of resources $n < N$.

PROPOSITION 7.5.– *The number of resources R that are simultaneously busy with offered traffic A follows an Engset distribution expressed by:*

$$\mathbb{P}(R = k) = \frac{\binom{N}{k}\left(\frac{A}{N-A}\right)^k}{\sum_{i=0}^{n}\binom{N}{i}\left(\frac{A}{N-A}\right)^i} \qquad [7.22]$$

PROOF.–

Again, as we did with Proposition 6.6 in section 6.2.2:

$$\pi_k = \binom{N}{k}(ch)^k \pi_0$$

where $ch = A/(N-A)$. The states that the system can take are $k = 0, 1, \dots, n$, so we have:

$$\frac{1}{\pi_0} = \sum_{i=0}^{n}\binom{N}{k}(ch)^i = \sum_{i=0}^{n}\binom{N}{k}\left(\frac{A}{N-A}\right)^i$$

Thus,

$$\pi_k = \frac{\binom{N}{k}(ch)^k}{\sum_{i=0}^{n}\binom{N}{i}\left(\frac{A}{N-A}\right)^i} = \frac{\binom{N}{k}\left(\frac{A}{N-A}\right)^k}{\sum_{i=0}^{n}\binom{N}{i}\left(\frac{A}{N-A}\right)^i} \qquad ■$$

The distribution we arrive at is called an *Engset distribution* or a *truncated binomial Erlang distribution*, or simply a *truncated binomial distribution*.

The term $b = A/(N-A)$ is the offered traffic per available source, since we have the offered traffic A, A busy sources, as well as $N-A$ free sources.

The blocking probability is obtained when $k = n$:

$$B = B_{n,N}(b) = \frac{\binom{N}{n}\left(\frac{A}{N-A}\right)^n}{\sum_{i=0}^{n}\binom{N}{i}\left(\frac{A}{N-A}\right)^i} = \frac{\binom{N}{n}b^n}{\sum_{i=0}^{n}\binom{N}{i}b^i} \qquad [7.23]$$

The loss probability is equal to the proportion of the number of refused requests to the total number of requests during a unit of time. The probability of a request when the system is in state k (k busy resources) is equal to $\mathbb{P}(R=k).(N-k)c$. Thus, the loss probability is:

$$E_{n,N}(b) = \frac{\mathbb{P}(R=n).(N-n)c}{\sum_{k=0}^{n} \mathbb{P}(R=k).(N-k)c} = \frac{\binom{N}{n} b^n (N-n)}{\sum_{k=0}^{n} \binom{N}{k} b^k.(N-k)}$$

$$= \frac{\binom{N-1}{n} b^n}{\sum_{k=0}^{n} \binom{N-1}{k} b^k} = E_{n,N-1}(b)$$

$$E_{n,N}(b) = B_{n,N-1}(b) \qquad [7.24]$$

From equation [7.24], we can say that the probability that a request coming from a random source will be rejected is equal to the probability that all $N-1$ remaining sources will keep the n resources busy.

The loss probability is markedly inferior to the blocking probability:

$$E_{n,N}(b) < E_{n,N+1}(b) = B_{n,N}(b) \qquad [7.25]$$

The carried traffic is equal to:

$$Y = A.\left(1 - B_{n,N-1}(b)\right) \qquad [7.26]$$

The lost traffic is equal to:

$$A_p = A.B_{n,N-1}(b) \qquad [7.27]$$

The improvement factor gives the amount of additional carried traffic when the number of resources increases from n to $n+1$. Denoting the traffic carried Y_n if we have n resources, we obtain:

$$F_{n,N}(A) = Y_{n+1} - Y_n \qquad [7.28]$$

REMARK.– We can note that:

– when n is very large, the Engset distribution tends towards an Erlang distribution;

– for given A and n, the loss probability according to Engset is less than that which Erlang predicts. The difference is all the more marked when n is small;

– for $n = N$, the Engset distribution is reduced to a Bernoulli distribution.

7.3.3. *On the Engset loss formula*

From equations [7.23] and [7.24], we can draw the following conclusions:

– $E_{n,N}(b)$ and $B_{n,N}(b)$ depend on A, N and n. For dimensioning, we should instead solve for n in terms of E, N and A, which is not analytically possible, as with the Erlang-B formula. We can also determine n using a table or abacus;

– we can also demonstrate the recursive formula:

$$\frac{1}{B_{n,N}(b)} = 1 + \frac{n}{b.(N-n+1)} \cdot \frac{1}{B_{n-1,N}(b)}, \quad B_{0,N}(b) = 1 \qquad [7.29]$$

7.4. Imperfect loss systems

As mentioned in Definition 7.2, an imperfect loss system is a system for which having one available resource is insufficient for it to be accessible. The internal congestion of the system creates this limitation.

7.4.1. *Loss probability in an imperfect system with limited and constant accessibility*

We say that an imperfect loss system is in a blocked state if:

– it is in a state of congestion: all n resources are busy;

– k free resources among the n available ones are all inaccessible.

One solution proposed by Palm[3] and Jacobæus[4] consists of supposing the system is perfect. The blocking probability then follows an Erlang distribution. They found the loss probability of a system with constant accessibility to be the ratio between the loss probabilities when offering traffic A to a system with n resources and another with $n - k$ resources with the same traffic:

3 Conrad Conny Palm (1907–1951), Swedish electric engineer and statistician.

4 Anton Christian Jacobæus (1911–1988), Swedish electric engineer.

$$E_k = \frac{\frac{A^n}{n!}}{\sum_{i=0}^{n}\frac{A^i}{i!}} \frac{\sum_{i=0}^{n-k}\frac{A^i}{i!}}{\frac{A^{n-k}}{(n-k)!}} = \frac{B_{1,n}(A)}{B_{1,n-k}(A)} \quad [7.30]$$

Equation [7.30] determines the probability of having k resources in activity among the n. The $n-k$ other resources are in another state.

Another solution, called the modified Palm–Jacobæus equation, also resolves this blockage problem in an imperfect system. The issue with the previous hypothesis is that it supposes an Erlang distribution of resource activity. To correct this incoherence, without however losing the advantage offered by the hypothesis of an Erlang distribution, we use fictitious offered traffic $\tilde{A}$ such that:

– an imperfect system (with n servers and accessibility k) offers traffic $\tilde{A}$;

– a perfect system (with n servers) offers fictitious corrected traffic.

– both carry the same traffic. The modified Palm–Jacobæus equation is given by:

$$E_k = \frac{B_{1,n}(\tilde{A})}{B_{1,n-k}(\tilde{A})} \quad [7.31]$$

This equation can only be resolved with iterations.

7.4.2. *Losses in a system with limited and variable accessibility*

The variation of accessibility in an imperfect loss system with variable accessibility cannot be modeled simply. We consequently do not have a simple calculation method for estimating losses.

To nevertheless determine the losses of such a system, we can resort to either the theory of graphs or to rudimentary simulations on a computer. We can also use the average accessibility offered by the system with the Palm–Jacobæus methods described in section 7.4.1. There are many other methods also based on this notion of average accessibility.

7.5. Exercises

1) Why do we prefer to use large beams rather than many smaller beams? What is the primary inconvenience of this way of doing things?

2) Imagine a scenario where the Engset equation must be used rather than the Erlang equation and justify your reasoning.

3) In which cases is a system called imperfect?

4) Imagine a system concentrating N lines out of n lines, with $N > n$. Can this system be perfect? Can we make it an imperfect system?

EXERCISE 7.1.– *PABX*

A company has 200 PABX switchboards. The traffic per switchboard is estimated to be 0.035 Erl departing and 0.045 Erl arriving.

1) Calculate the total traffic.

2) What is the loss probability if the company switch only has 30 circuits?

EXERCISE 7.2.– *Erlang-B table*

A loss system has M circuits.

1) What is the required traffic for a rate of loss of 1%, 10% and 50% when M is, respectively, equal to 5, 10 and 15?

2) For each value, determine the traffic that is offered, carried and lost (use the Erlang-B table or the Erlang-B formulas).

EXERCISE 7.3.– *GSM cell*

A GSM cell has 18 circuits. Traffic is offered to these circuits such that the call rate is 480 calls/hour and the average duration of calls is 105 seconds.

1) What are the offered traffic, the blocking rate, the loss rate, the total carried traffic and carried traffic per circuit?

2) What is the expected number of rejected calls per hour?

EXERCISE 7.4.– *Authorized traffic and yield*

Two switch systems are connected by two lines with 10 circuits each. Suppose a rate of loss of 5%.

1) What are the authorized traffic per line and the yield per line?

2) What is the total authorized traffic for the two lines?

We regroup the two lines into one with 20 circuits.

3) Supposing the same rate of loss, what are the new authorized traffic and yield?

EXERCISE 7.5.– *Line for outgoing calls*

A company has a call service center with 120 phones, of which only 100 have access to the outside. Suppose that a normal user has a telephone traffic of 0.12 Erl divided as follows:

– 0.04 Erl in outgoing traffic;

– 0.04 Erl in incoming traffic;

– 0.04 Erl in internal traffic within the company.

Define:

1) the total switch capacity in Erl;

2) the line for outgoing calls, given that when the line is busy, callers get a busy signal (the failure rate must not go above 10%).

EXERCISE 7.6.– *Total accessibility group (1)*

A group (of resources) with total accessibility is made up of ten resources and is offered traffic by ten sources. Suppose the offered traffic per source is equal to 0.5 Erl.

Calculate the offered traffic, carried traffic, lost traffic, the blocking rate and the loss rate.

EXERCISE 7.7.– *Total accessibility group (2)*

A group (of resources) with total accessibility is made up of ten resources and is offered traffic by a large number of sources. Suppose the total offered traffic is equal to 5 Erl.

Calculate the offered traffic, carried traffic, lost traffic, the blocking rate and the loss rate.

EXERCISE 7.8.– *Call center*

A call center estimates that, for an upcoming television show, they will need to receive approximately 720,000 calls in 2 hours. Supposing the average duration of a call is 30 seconds, what are the number of necessary lines and the carried traffic in Erl?

EXERCISE 7.9.– *Back to the past of the X.25*

A catalogue sales company, besides Internet access, continues to offer its customers an ordering and browsing service with its catalogue on Minitel. Given the reduction in activity of this service, the company must redimension its needs. The information to consider is: a transaction for orders lasts, on average, 2 minutes, and at the peak hour, there are 600 connection requests. To completely satisfy its customer, the company would like the refusal rate due to a lack of resources to be no greater than 1%. The videotex service is accessible on an X.25 network and accessible via the switch telephone network.

1) How many virtual circuits must the company subscribe to from the X.25 operator?

2) In the previously established conditions, how many connection requests will be possible in 1 hour, if we suppose a refusal rate of 2%?

8

Teletraffic in Delay Systems

It's the time spent in waiting rooms that turns sick people into patients.

Claude Frisoni (1954–)

In this chapter, we study the teletraffic of systems in which requests that cannot be processed immediately due to insufficient available resources are put on hold. They are processed one by one as resources for processing become available.

8.1. Delay system

8.1.1. *Description*

Let us consider a delay system with total accessibility: when all the resources of the system are busy, requests arriving from sources join a queue and wait there until a resource is free. No request should be asked to wait if a resource is free; in other words, all requests will be processed. There are *no lost requests*.

In a delay system, lost traffic or rejected traffic A_p is equal to 0. Carried traffic is therefore equal to offered traffic. However, the notion of the *maximum capacity of spendable traffic* must be posited to avoid infinite delay times.

With type-one pure chance traffic, the arrival of requests from an infinite number of sources is Poissonian and the duration of activity is distributed according to an exponential distribution. This model is called an Erlang delay system model. The distribution of delay times in such a system can be calculated according to the way requests are organized in the waiting room: FIFO, LIFO, random, etc.

With type-two pure chance traffic, requests arrive from a limited number of sources, and the duration of activity is distributed according to an exponential distribution.

This model is called the Palm model or Palm's machine-repair model. It is often used for estimating dimensions of computer systems.

8.1.2. *Characteristics of delay*

The main criterion of a delay system is the waiting time or delay. This delay can be discrete or continuous depending on whether or not the service process is measured by the clock.

DEFINITION 8.1.– *The delay, denoted d, is the lapse of time between the arrival of a request and the start of resource activity in response to this request.*

The delay d is a random variable. It can equal zero in the case where at least one resource is free at the instant a request arrives, that is, a resource is immediately attributed to it.

The delay depends on the dimensions of the space where waiting requests are stored, called the *queue*. We use q to denote this dimension of the queue.

We have:

– $\mathbb{P}(d > t)$ the probability of surpassing a certain delay time t;

– $\mathbb{P}(d > 0) = D$ the probability of waiting. D can be interpreted as the proportion of delayed requests, and $1 - D$ the proportion of immediately accepted requests;

– the average delay time of delayed requests, without taking immediately accepted requests into account:

$$E(d_q) = \frac{E(d)}{\mathbb{P}(d > 0)} = \frac{E(d)}{D} \quad [8.1]$$

Contrary to loss systems, in delay systems, the distribution of activity durations h influences the distribution of delay times d. We can distinguish the following two borderline cases:

– the previously described pure chance traffic where activity duration h is distributed according to an exponential distribution. The density function of h is equal to:

$$f(h) = \frac{1}{E(h)} e^{-\frac{h}{E(h)}} \quad [8.2]$$

– a constant activity duration $h = E(h) = constant$. This is a totally deterministic process, as is frequently found in online orders.

In practice, we rarely encounter these two borderline cases. However, they provide us with two values that act as a range for estimating general cases of activity duration h, which are generally difficult to express mathematically.

8.2. Erlang model

In this section, we consider a delay system with n resources and total accessibility whose waiting capacity is infinite. The number of sources is infinite and the traffic taken to be type-one pure chance traffic.

8.2.1. *Infinitely long queue*

Since the traffic is type-one pure chance traffic, arrivals to the system follow a Poisson process with intensity $\lambda = c$. For an activity duration distributed according to an exponential distribution whose average is $\mu = 1/h$, the system can be modeled by an M/M/n queue (see the description of this queue in section 3.6).

This model is called the Erlang model for a delay system.

According to Proposition 3.9, replacing ρ by $A = \lambda/\mu = c.h$, the stationary distribution of the M/M/n queue is equal to:

$$\begin{aligned} \pi_k &= \frac{\frac{A^k}{k!}}{\frac{A^n}{n!} \cdot \frac{n}{n-A} + \sum_{i=0}^{n-1} \frac{A^i}{i!}} && \text{(for } k \leq n\text{)} \\ \pi_k &= \frac{\frac{A^k}{n!\, n^{k-n}}}{\frac{A^n}{n!} \cdot \frac{n}{n-A} + \sum_{i=0}^{n-1} \frac{A^i}{i!}} = \pi_n \cdot \left(\frac{A}{n}\right)^{k-n} && \text{(for } k \geq n\text{)} \end{aligned} \quad [8.3]$$

Due to the ergodicity, the probability that a request from some source will wait before being processed is equal to the proportion of time that all the resources are busy.

We thus have, given that $A/n < 1$:

$$\mathbb{P}(d > 0) = E_{2,n}(A) = \sum_{k=n}^{+\infty} \pi_k = \pi_n . \frac{n}{n-A}$$
$$= \frac{\frac{A^n}{n!}\frac{n}{n-A}}{1 + A + \frac{A^2}{2!} + \cdots + \frac{A^{n-1}}{(n-1)!} + \frac{A^n}{n!} \cdot \frac{n}{n-A}} \quad [8.4]$$

This equation [8.4] is called the *Erlang-C formula* or the *second Erlang formula.*

The probability that a request will be immediately accepted is equal to $1 - D = 1 - E_{2,n}(A)$.

Denoting the random variable for the length of the queue L, the probability of finding a request waiting on hold at any instant is equal to:

$$\mathbb{P}(L > 0) = \sum_{k=n+1}^{+\infty} \pi_k = \sum_{k=n+1}^{+\infty} \frac{\frac{A^k}{n!\, n^{k-n}}}{\frac{A^n}{n!} \cdot \frac{n}{n-A} + \sum_{i=0}^{n-1} \frac{A^i}{i!}}$$
$$= \frac{\frac{A}{n}}{1 - \frac{A}{n}} \pi_n = \frac{A}{n-A} \pi_n = \frac{A}{n} E_{2,n}(A) \quad [8.5]$$

8.2.2. *Erlang-C formula*

The formula from equation [8.4] is called the Erlang-C formula. It designates the probability D that a request will be delayed.

$D = B_{2,n}(A)$ depends on A and n. For dimensioning, we ought to calculate n with respect to D and A, which is not analytically possible. The determination of n is undertaken using a table called the *Erlang-C table*, provided in Appendix 2, or using an abacus, as in Figure 8.1. An estimate from approximation formulas can also be useful in certain cases.

The Erlang-C formula is similar to the Erlang-B formula except for the factor $n/(n-A)$ in the last term. So, if we have $B_{1,n}(A)$, we can calculate $B_{2,n}(A)$ using the formula:

$$B_{2,n}(A) = \frac{n.B_{1,n}(A)}{n - A.\left(1 - B_{1,n}(A)\right)} = \frac{B_{1,n}(A)}{1 - \frac{A}{n}.\left(1 - B_{1,n}(A)\right)} \qquad [8.6]$$

In equation [8.6], note that $B_{2,n}(A) \geq B_{1,n}(A)$ for any given n and A.

We can also demonstrate the following recursive formula, found by Sanders[1]:

$$\frac{1}{B_{2,n}(A)} = \frac{1}{B_{1,n}(A)} - \frac{1}{B_{1,n-1}(A)} \qquad [8.7]$$

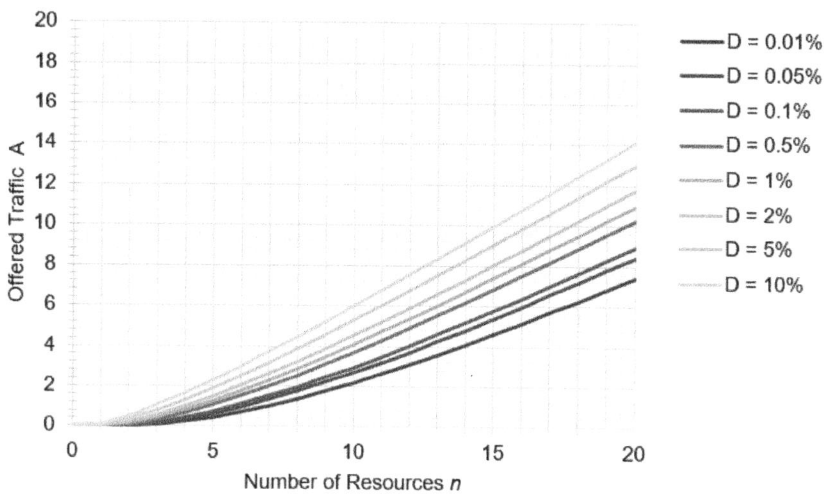

Figure 8.1. *Abacus for Erlang-C formula. For a color version of this figure, see www.iste.co.uk/rava/queues.zip*

8.2.3. *Distribution of delays*

8.2.3.1. *Number of waiting requests*

PROPOSITION 8.1.– *The average number $E(N_q)$ of requests that are waiting at any given instant is equal to:*

1 B. Sanders, Dutch university student.

$$E(N_q) = D.\frac{A}{n-A} \qquad [8.8]$$

PROOF.–

We have 0 wait when $k = 0, 1, \ldots, n$ requests can be simultaneously processed by the system, and $k - n$ waiting requests when $k \geq n$ requests arrive in the system.

Thus:

$$E(N_q) = \sum_{k=0}^{n} 0.\pi_k + \sum_{k=n+1}^{+\infty} (k-n).\pi_k$$

$$= \sum_{k=n+1}^{+\infty} (k-n).\left(\frac{A}{n}\right)^{k-n}.\pi_n = \pi_n.\sum_{k=n+1}^{+\infty} (k-n).\left(\frac{A}{n}\right)^{k-n}$$

$$= \pi_n.\sum_{k=1}^{+\infty} k.\left(\frac{A}{n}\right)^{k} = \pi_n.\frac{\frac{A}{n}}{\left(1-\frac{A}{n}\right)^2} = \pi_n.\left(\frac{n}{n-A}\right)\left(\frac{A}{n-A}\right)$$

$$= D.\frac{A}{n-A} \quad \blacksquare$$

PROPOSITION 8.2.– *The average number $E(N_{qa})$ of requests that are waiting when there is a queue is equal to:*

$$E(N_{qa}) = \frac{n}{n-A} \qquad [8.9]$$

PROOF.–

The conditional expected value is given by:

$$E(N_{qa}) = \frac{\sum_{k=n+1}^{+\infty}(k-n).\pi_k}{\sum_{k=n+1}^{+\infty}\pi_k} = \frac{\pi_n.\left(\frac{n}{n-A}\right)\left(\frac{A}{n-A}\right)}{\pi_n.\left(\frac{A}{n-A}\right)} = \frac{n}{n-A} \quad \blacksquare$$

8.2.3.2. *Delay times*

We can distinguish two types of delays:

– the delay time for all requests, denoted d, with its expected value denoted $W = E(d)$;

– the delay time for requests that end up having to wait, denoted d_q, with its expected value denoted $w = E(d_q)$.

PROPOSITION 8.3.– *The average waiting time W for all requests is equal to:*

$$W = E(d) = D.\frac{h}{n - A} \quad [8.10]$$

where h designates the average duration of activity.

PROOF.–

Using the Little formula from equation [3.9] for the average length of the queue itself $E(N_f)$ and the waiting time, we have:

$$E(N_q) = \lambda . E(d) = cW$$

Thus:

$$W = E(d) = \frac{1}{c}L_n = \frac{1}{c}D.\frac{A}{n - A} = \frac{1}{c}D.\frac{ch}{n - A}$$
$$= D.\frac{h}{n - A} \quad ■$$

PROPOSITION 8.4.– *The average waiting time w for requests that end up having to wait is equal to:*

$$w = E(d_q) = \frac{h}{n - A} \quad [8.11]$$

PROOF.–

We simply need equation [8.1]: $w = W/D$. ■

The distribution of delay times d can be calculated when we note that a particular request advances by one position in the queue each time a resource becomes free. This release happens with a probability of $nhdt$ for an interval of time dt. It is the death rate for the Markov chain that corresponds to the system model for a state $x \geq n$. The interval of time between two shifts in the chain is distributed exponentially with an average value of nh. The request occupying state i in the chain will undergo i shifts before being processed, that is, the sum i of random variables distributed according to the same exponential distribution. The result is a gamma distribution with parameters nh and i. Taking all possible values of i (from 1 to infinity) into account, we finally arrive at the probability that d will be greater than a given amount of time:

$$\mathbb{P}(d > t) = D.e^{-\frac{t}{E(d_q)}} = D.e^{-\frac{t(n-A)}{h}} \quad [8.12]$$

Delay times are therefore exponentially distributed, just like the distribution of activity duration.

REMARK.– In the results obtained up to this point, we took the queue's FIFO character into account. The service strategy has a direct influence on the delay times, which is logical. Thus, for a different organizational strategy regarding queues, the probability of having long delays will be different from the exponential distribution.

8.3. Finite waiting capacity model

Creating a system with an infinite waiting capacity is practically impossible. A more realistic model of a system with a finite waiting capacity is considered in this section. Once this waiting capacity is reached, a newly arriving request will be rejected. The system is therefore considered a *loss delay system*.

Let us consider type-one pure chance traffic. The distribution of service duration is exponential.

We denote the waiting capacity of the system q, that is, the maximum number of requests that can wait is equal to q.

8.3.1. *Queues of finite length*

According to these hypotheses, the system can be modeled by an M/M/n/S queue such that $S = n + q$. This queue can in turn be modeled by a Markov chain with $S + 1$ states. Let π_k be its stationary distribution.

The loss probability of the system can be found when $k = S = n + q$:

$$B = \pi_{n+q} \quad [8.13]$$

The waiting probability of the system can be found when $k \geq n$:

$$D = \sum_{k=n}^{n+q-1} \pi_k \quad [8.14]$$

The carried traffic is equal to the average number of busy resources, so it can be expressed as:

$$Y = \sum_{k=0}^{n} k.\pi_k + \sum_{k=n+1}^{n+q} n.\pi_k \qquad [8.15]$$

In practice, dimensioning such a system consists of determining the waiting capacity q from an admissible loss probability B.

The calculation of the stationary is left to the reader as an exercise. Given the complexity of the analytic expression of the result, it is difficult to analytically determine what expression is needed when calculating the waiting capacity q. We must therefore use the infinite capacity model while also accounting for a loss B.

8.3.2. *Limitations affecting the delay*

In certain realistic cases, we find impatient requests. These types of requests do not tolerate significant delay times.

We can consider, for a particular system with an exponential distribution of delay times, that requests whose delays exceed a threshold d_0 are abandoned by the source.

We therefore have a loss probability (due to the sources who cancel requests and not because of insufficient resources or waiting capacity) of:

$$B_d = \mathbb{P}(d > d_0) = d.e^{-\frac{t}{E(d_q)}} \qquad [8.16]$$

These cancellations modify the state of the queue and the death rate of the corresponding Markov chain for $x > n$. Equations [8.3] to [8.11] are therefore not applicable to the calculation of this loss probability beyond an approximation.

8.4. Palm model

Let us consider a delay system with n resources and total accessibility. This time, we consider the traffic to be by type-two pure chance, that is, coming from a finite number of sources. The number of sources offering the offered traffic to the system is denoted by N. An arriving request is accepted for a service if a system resource is free; it is put on hold if all the resources are busy (no loss).

8.4.1. *M/M/n/N/N queue*

With type-two pure chance traffic, the process of arrivals is no longer Poissonian. According to section 6.2, by modeling the system with a birth and death process whose state is the number of accepted requests for a service k or put on hold, the growth rate of the corresponding Markov chain is equal to $\lambda_k = c.(N-k)$ (see the hypotheses in section 2.3.2).

The durations of activity are exponential with an average of $1/\mu = h$. We thus have an M/M/n/N/N queue, where N designates the number of sources and n is the number of resources. In everything that follows, we consider the system to have an insufficient number of resources for requests that are not immediately accepted to be put on hold ($n < N$).

Figure 8.2 represents the Markov chain modeling this system.

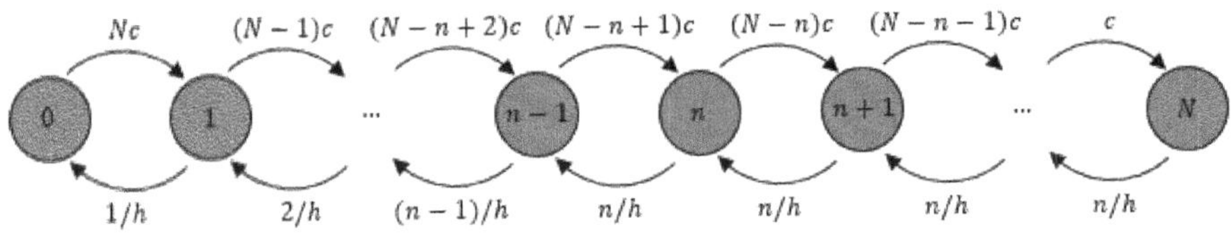

Figure 8.2. *Markov chain for the Palm model*

The stationary distribution of this model is equal to:

$$\pi_k = \binom{N}{k}(ch)^k \pi_0 \qquad for\ 0 \le k \le n$$

$$\pi_k = \frac{(N-n)!}{(N-k)!}\left(\frac{ch}{n}\right)^{k-n} \pi_n \qquad for\ n \le k \le N \tag{8.17}$$

By solving this stationary distribution using $\sum_{k=0}^{N} \pi_k = 1$, we find that:

For $0 \le k \le n$:

$$\pi_k = \frac{\binom{N}{k}(ch)^k}{\sum_{i=0}^{n}\binom{N}{i}(ch)^i + \sum_{i=n+1}^{N}\frac{N!}{(N-i)!\,n!\,n^{i-n}}(ch)^i} \tag{8.18}$$

For $n \le k \le N$:

$$\pi_k = \frac{\frac{N!}{(N-k)!\,n!\,n^{k-n}}(ch)^k}{\sum_{i=0}^{n}\binom{N}{i}(ch)^i + \sum_{i=n+1}^{N}\frac{N!}{(N-i)!\,n!\,n^{i-n}}(ch)^i} \quad [8.19]$$

Recall that $ch = A/(N-A)$.

8.4.2. *Characteristics of traffic*

Let us hypothetically say that the loss probability is equal to $B = 0$.

The probability of waiting D is equal to π_n:

$$D = \frac{\left(\frac{A}{N-A}\right)^N}{\sum_{i=0}^{n}\binom{N}{i}\left(\frac{A}{N-A}\right)^i + \sum_{i=n+1}^{N}\frac{N!}{(N-i)!\,n!\,n^{i-n}}\left(\frac{A}{N-A}\right)^i} \quad [8.20]$$

For the particular case where $n = 1$, we have:

$$D = \frac{\left(\frac{A}{N-A}\right)^N}{\sum_{i=0}^{N}\frac{N!}{(N-i)!}\left(\frac{A}{N-A}\right)^i} \quad [8.21]$$

The carried traffic is equal to the offered traffic $Y = A$.

The average number of delayed requests and the waiting times are left to the reader as an exercise.

8.5. General distribution model for activity

8.5.1. *The Pollaczek–Khinchine formula*

Exponential modeling of activity durations was set out to simplify mathematical calculations but, in general, these durations do not follow an exponential distribution. We here consider a system with only one resource whose activity durations are independent and have a general distribution. This system can be modeled by an M/GI/1 queue, already discussed in section 3.7.

The Pollaczek–Khinchine formula [3.34] of an M/GI/1 queue allowed us to arrive at equation [3.35], provided here:

$$E(N_S) = \rho + \frac{\rho^2(1 + C_S^2)}{2(1 - \rho)} \qquad [8.22]$$

where S designates the random variable of the number of customers in the M/GI/1 delay system; ρ is the offered traffic, which we denote A in everything that follows; and $C_S^2 = \sigma^2(S)/E^2(S) = \sigma^2/h^2$.

The average number of requests waiting in the queue is thus equal to (see Exercise 3.1, recalling that $\lambda_e/\mu = \rho$):

$$E(N_q) = \frac{A^2(1 + C_S^2)}{2(1 - A)} \qquad [8.23]$$

From equation [8.23], we can obtain:

– the average delay time (following the Little formula):

$$W = E(d) = \frac{E(N_q)}{c} = \frac{c(h^2 + \sigma^2)}{2(1 - A)} \qquad [8.24]$$

– the average waiting time for delayed requests:

$$w = E(d_q) = \frac{h^2 + \sigma^2}{2h(1 - A)} \qquad [8.25]$$

8.5.2. *Activity with a constant duration*

A particular case of the distribution of activity durations is constant duration. We denote $E(h) = h$ constant, and we have $\sigma^2 = 0$.

The average number of waiting requests is equal to:

$$E(N_q) = \frac{A^2}{2(1 - A)} \qquad [8.26]$$

The average delay time is equal to:

$$W = E(d) = \frac{hA}{2(1-A)} \quad [8.27]$$

The average waiting time of delayed requests is equal to:

$$w = E(d_q) = \frac{h}{2(1-A)} \quad [8.28]$$

Compared to equation [8.11] with an exponential distribution of activity durations, the average waiting time of delayed requests is two times shorter for identical traffic with the same activity average h.

Equation [8.28] only indicates the average value of the delay, but its distribution is not exponential as with the case of the exponential distribution of durations.

8.6. Exercises

1) What is the loss probability of a delay system?

2) What are the determining quantities of a delay system?

3) Is it better to define several queues or one general queue? What do we gain by choosing one solution rather than the other?

4) Cite an example of a delay system in telecommunications (not the queue at the post office counter please).

EXERCISE 8.1.– *Total accessibility with delay (1)*

Let there be a group of lines with total accessibility in a delay system made of 30 circuits. It is offered traffic with a call rate of 700 calls/hour. The average duration of calls is 108 seconds.

Suppose that the Erlang distribution hypotheses for delay systems are fulfilled and that calls are processed according to their order of arrival.

Calculate the offered traffic, the probability of delay, the carried traffic, the traffic carried per channel, the average delay for all calls and the average delay for calls that must wait.

EXERCISE 8.2.– *Total accessibility with delay (2)*

Let there be a group with total accessibility in a delay system made of 30 circuits. It is offered traffic with a call rate of 700 calls/hour. The average duration of calls is 108 seconds.

Suppose that the Erlang distribution hypotheses for delay systems are fulfilled and that calls are processed according to their order of arrival.

1) Calculate the probability that a call will have to wait more than 3 seconds, 6 seconds or 12 seconds.

2) Calculate the probability that a call will have to wait more than 3 seconds, 6 seconds or 12 seconds, given that it must wait.

EXERCISE 8.3.– *Phone-in*

A radio show phone-in lasts one hour. During the phone-in, calls are meant to join the program's delay system, which follows a Poisson process including an average of 30 calls during this hour-long show.

The lengths of time that calls from different listeners last are considered independent of each other and of the arrival process. Suppose they follow an exponential distribution whose average is equal to h seconds.

Suppose also that no listeners already waiting will be lost and that calls in the queue are served in the order of their arrival.

What is the greatest allowable value for h in seconds that will satisfy the condition that the probability of a call having to wait more than 3 seconds will not be greater than 0.1?

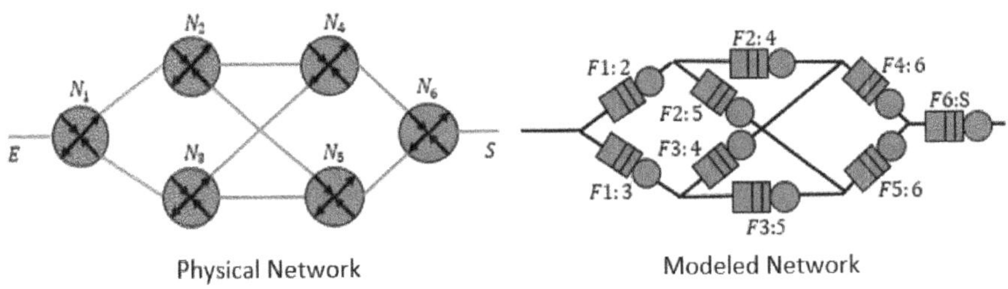

Figure 8.3. *Network from Exercise 8.3*

EXERCISE 8.4.– *Wait*

Let us consider an Erlang delay system with 10 circuits and an offered traffic of 8 Erl. The average duration of activity is 5 minutes. Waiting calls are taken in the order received.

What is the probability that an incoming call will encounter another call in the queue for more than 30 seconds?

EXERCISE 8.5.– *Line for incoming calls*

A company has a call service center with 120 phones, of which only 100 have access to the outside. Suppose a normal user has telephone traffic of 0.12 Erl, distributed as follows:

– 0.04 Erl in outgoing traffic;

– 0.04 Erl in incoming traffic;

– 0.04 Erl in internal traffic within the company.

Define the incoming call line given that callers hear music while waiting when the line called is busy (the failure rate should not surpass 2%).

EXERCISE 8.6.– *Constant holding*

Requests arrive at a data server that is similar to a delay system and follow a Poisson process with a rate of 4,320 requests per minute. The processing time of the server is taken to be constant and is equal to 3 microseconds.

The processing time and the inter-arrivals of requests are independent, and no requests already in the queue leave it.

1) What is the probability that an arriving request will have to wait?

2) What is the average delay time for requests that have to wait?

3) What is the average delay time for all of the requests?

4) What is the average response time of the server?

EXERCISE 8.7.– *Terminals in a distribution chain*

An appliance distribution chain wishes to construct a new retail space. The new store will have passive terminals connected to the central computing system of the company. The computer applications are of four kinds:

– Customer reception and sales: terminals (point of sales terminals) allow salesclerks to inform customers about the availability of a product, with the service node being in use for approximately one minute. When the customer acquires the object (6 times out of 10, on average), taking the order and updating the stock takes approximately another 3 minutes, during which the service node is unavailable to others.

– The collections counter has terminals that are used for identifying the exact location of purchased objects to be collected: the counter is in use for approximately one minute per collection (each sale will result in a collection).

– Payments: it is expected that there will never be more than five customers waiting in front of a cash register and that a customer becomes impatient if required to wait more than 10 minutes before being served; the payment process takes on average 3 minutes.

– The system also has two accountant stations (one per agent). Each accountant makes at most 20 transactions/day, regularly distributed throughout the day, and works 8 hours/day. Let us suppose that at peak hour, on the busiest day, 100 customers on average are helped by the salesclerks.

Define:

1) the number of register terminals;

2) the number of purchase collection counters, if we accept that in 80% of cases, store clerks must find an available terminal;

3) the number of terminals available to salesclerks, given that in 95% of cases, the sales clerk should find an available one;

4) given the characteristics of the applications provided by Table 8.1, determine the response time of the application.

The different factors to consider for this calculation are:

– the processing time for a transaction by the central server is 0.2 s;

– the terminals are connected to a local hub by a line 9,600 bit/s;

– the store is connected to the central site by a rented connection of 64,000 bit/s;

– we can assume the transparency of binary bits and the service data (ACK, etc.) increase the dimensions of units by 20%;

– the screen interface at the sales terminals is organized such that it allows for the sale of the viewed product or for another product to be searched (only one exchange per page followed by a sale);

– entries take place on a new window or over the top of another response window; only the characters entered are transmitted;

– the entry time is not counted in the response time;

– the screen loading time is considered negligible.

Terminal	Sales point	Register	Collection Point	Accounting
Entry	20c	100c	20c	200c
Response or new entry page	800c	500c	600c	800c

Table 8.1. *Transaction characteristics*

EXERCISE 8.8.– *Teletraffic in a local area network*

A local area network is connected to another network via a router with a line of 64 kbit/s. Numerous computers are connected on the local area network. Arriving traffic analysis shows that:

– two computers have an outgoing traffic of four packets/s;

– two computers have an outgoing traffic of two packets/s;

– three computers have an outgoing traffic of six packets/s;

– five computers have an outgoing traffic of five packets/s.

The arrivals follow a Poisson distribution. The packets, when they arrive, have an average length of 128 bytes. We will ignore protocol data.

Given these details, determine:

1) the arrival rate (λ);

2) the service rate of the router (μ);

3) the traffic intensity or the system load (ρ);

4) the average number of packets in the router;

5) the average delay time;

6) the average number of packets waiting;

7) the response time;

8) the size of the entry buffer if dimensioned as exactly as possible for this traffic, rounding up to the next kilobyte;

9) with the size of the buffer no longer being infinite, in these conditions, what is the probability of a new entry being rejected?

EXERCISE 8.9.– *Teletraffic in a computer network*

The network in Figure 8.2 is made of connections of 64 kbit/s; it uses random routing, and traffic analyses show that the entering traffic per node E is on average 30 packets per second and of an average length of 128 bytes. Let us assume there is no other source of traffic in the network. All traffic entering E leaves via S and is statistically spread as shown in Figure 8.3. Determine the average transit time of one packet in the network.

Connection	Proportion of carried traffic
$N_1 - N_2$	75%
$N_2 - N_4$	50%
$N_3 - N_5$	25%

Table 8.2. *Proportion of traffic carried in connections*

PART 4

Answers to Exercises

9

Chapter 1 Exercises

EXERCISE 1.1.– *Radioactive particle*

We use T to denote the random variable equal to the lifetime of the radioactive particle.

T follows an exponential distribution with parameter λ, so its expected lifetime is equal to $E(T) = 1/\lambda$.

The probability that it will still be alive after its expected lifetime is equal to $\mathbb{P}(T > 1/\lambda) = e^{-\lambda.\frac{1}{\lambda}} = e^{-1}$.

Its expected lifetime given that the particle is still alive after its expected lifetime is equal to $E(T \mid T > 1/\lambda)$. Its value is still equal to $1/\lambda$ given the amnesia property of the exponential distribution.

We can also demonstrate this in another way. We first determine the probability distribution for the variable $T \mid T > 1/\lambda$.

$$\mathbb{P}(T > t|T > 1/\lambda) = \frac{\mathbb{P}(T > t, T > 1/\lambda)}{\mathbb{P}(T > 1/\lambda)}$$

Now, we can write that $\mathbb{P}(T > t, T > 1/\lambda) = \mathbb{P}(T > t + 1/\lambda)$ since the exponential distribution is memoryless: $\mathbb{P}(T > t, T > 1/\lambda) = \mathbb{P}(T > t + 1/\lambda) = e^{-\lambda\left(t+\frac{1}{\lambda}\right)}$.

$$\mathbb{P}(T > t|T > 1/\lambda) = \frac{\mathbb{P}(T > t, T > 1/\lambda)}{\mathbb{P}(T > 1/\lambda)} = \frac{e^{-\lambda\left(t+\frac{1}{\lambda}\right)}}{e^{-1}} = e^{-\lambda t}$$

$T \mid T > 1/\lambda$ again follows an exponential distribution with parameter λ. Its expected value is thus $1/\lambda$.

EXERCISE 1.2.– *Discretization of exponential distribution*

X is a random exponential variable with parameter λ.

We use S to denote the ceiling function of X: $S = \lceil X \rceil = n$ is equivalent to $n - 1 < X \leq n,\ \forall n \in \mathbb{N}^*$. The probability distribution of S is therefore:

$$\mathbb{P}(S = n) = \mathbb{P}(n - 1 < X \leq n) = \int_{n-1}^{n} \lambda e^{-\lambda t} dt$$

$$\mathbb{P}(S = n) = \int_{n-1}^{n} \lambda e^{-\lambda t} dt = e^{-\lambda(n-1)} - e^{-\lambda n} = e^{-\lambda(n-1)}\left(1 - e^{-\lambda}\right)$$
$$= \left(1 - e^{-\lambda}\right)\left(1 - \left(1 - e^{-\lambda}\right)\right)^{n-1}$$

Thus, $S = \lceil X \rceil$ follows a geometric distribution with parameter $\left(1 - e^{-\lambda}\right)$.

We use I to denote the floor function of X: $I = \lfloor X \rfloor = n$ is equivalent to $n \leq X < n + 1,\ \forall n \in \mathbb{N}$. The probability distribution of I is therefore:

$$\mathbb{P}(I = n) = \mathbb{P}(n \leq X < n + 1) = \int_{n}^{n+1} \lambda e^{-\lambda t} dt$$

$$\mathbb{P}(I = n) = \int_{n}^{n+1} \lambda e^{-\lambda t} dt = e^{-\lambda n} - e^{-\lambda(n+1)} = e^{-\lambda n}\left(1 - e^{-\lambda}\right)$$
$$= \left(1 - e^{-\lambda}\right)\left(1 - \left(1 - e^{-\lambda}\right)\right)^{n}$$

Thus, $I = \lfloor X \rfloor$ also follows a geometric distribution with parameter $\left(1 - e^{-\lambda}\right)$. This is the second definition of the geometric distribution.

By proceeding with the same steps, we find that the distribution of the random variable ceiling function of X/τ (for any real number $\tau > 0$) is the geometric distribution with parameter $1 - e^{-\lambda\tau}$.

EXERCISE 1.3.– *Synchronization*

Let us denote $X = \min(X_1, X_2)$ and $\lambda = \lambda_1 + \lambda_2$.

By applying the equality of events $A_1 = \{X = X_1\}$ and $A_2 = \{X = X_2\}$, we find:

$$\begin{aligned}\mathbb{P}(X_1 = X_2) &= \mathbb{P}(A_1 \cap A_2) \\ &= \mathbb{P}(A_1) + \mathbb{P}(A_2) - \mathbb{P}(A_1 \cup A_2) \\ &= \frac{\lambda_1}{\lambda} + \frac{\lambda_2}{\lambda} - 1 = 0\end{aligned}$$

The variables X_1 and X_2 are thus equal with a probability of zero.

EXERCISE 1.4.– *"Very" random variable*

Entropy of the variable X_e:

$$\begin{aligned}h(X_e) &= -\int_0^{+\infty} \lambda e^{-\lambda t}.\ln\left(\lambda e^{-\lambda t}\right) dt = -\int_0^{+\infty} \lambda e^{-\lambda t}.\ln(\lambda)\, dt + \int_0^{+\infty} \lambda e^{-\lambda t}.\lambda t dt \\ &= -(\ln \lambda)\int_0^{+\infty} \lambda e^{-\lambda t} dt + \lambda \int_0^{+\infty} \lambda t e^{-\lambda t} dt = -\ln\lambda + \lambda.\frac{1}{\lambda} \\ &= 1 - \ln\lambda\end{aligned}$$

Let $t \geq 0$. By using $f(t) = f_X(t)$ and $g(t) = \lambda e^{-\lambda t}$ over the Gibbs inequality, we have:

$$\int_0^{+\infty} f(t).\ln\left(\frac{f(t)}{g(t)}\right) dt = \int_0^{+\infty} f(t).\ln\big(f(t)\big)\, dt - \int_0^{+\infty} f(t).\ln\big(g(t)\big)\, dt$$

$$= \int_0^{+\infty} f_X(t).\ln\big(f_X(t)\big)\, dt - \int_0^{+\infty} f_X(t).\ln\left(\lambda e^{-\lambda t}\right) dt$$

$$= -h(X) - \int_0^{+\infty} f_X(t).\ln(\lambda)\, dt + \int_0^{+\infty} \lambda t f_X(t) dt \geq 0$$

Now $f_X(t)$ is a probability density function, so $\int_0^{+\infty} f_X(t)dt = 1$ and $\int_0^{+\infty} t f_X(t)dt = E(X) = 1/\lambda$.

$$-h(X) - \int_0^{+\infty} f_X(t).\ln(\lambda)\, dt + \int_0^{+\infty} \lambda t f_X(t) dt = -h(X) - \ln(\lambda) + 1 \geq 0$$

We thus find $h(X) \leq 1 - \ln\lambda$.

We have equality if and only if $f(t) = g(t)$, that is, $f_X(t) = \lambda e^{-\lambda t}$, where X is a random variable following an exponential distribution with parameter λ.

Thus, $h(X) \leq h(X_e)$ for any real positive random variable X with an average of $1/\lambda$. The exponential distribution is thus the law of maximal entropy among all the distributions with the same average and a density over $\mathbb{R}^+$.

EXERCISE 1.5.– *Call expiry*

We use $T_1 \rightsquigarrow \mathcal{E}(\lambda_1)$ to denote the duration of a personal call and $T_2 \rightsquigarrow \mathcal{E}(\lambda_2)$ to denote the duration of a professional call. $1/\lambda_1 = 0.5$ minutes is the average duration of a personal call, and $1/\lambda_2 = 1$ minute is that of a professional call. A completed call is equivalent to either a personal call or a professional call that finishes first, that is, the minimum between T_1 and T_2: $T = \min(T_1, T_2)$, the expiry time of the first call. T follows an exponential distribution with parameter $\lambda = \lambda_1 + \lambda_2 = 3$. We thus expect a completed call after $E(T) = \frac{1}{\lambda} = \frac{1}{3}$ minutes, or 20 seconds.

The probability that a professional call will be completed first is equal to $\lambda_2/\lambda = 1/3$.

At any instant t, we count n personal calls and m professional calls in progress at the same time: we use $T = \min(T_{11}, T_{12}, \dots, T_{1n}, T_{21}, \dots, T_{2m})$ to denote the expiry time of the first call. T_{1i} designates the duration of the i-th personal call, and T_{2j} designates that of the j-th call. T follows an exponential distribution with parameter $\lambda = n\lambda_1 + m\lambda_2$ and the probability that it will be a professional call that finishes first is equal to $\frac{\lambda_2}{\lambda} = \frac{\lambda_2}{n\lambda_1 + m\lambda_2}$.

EXERCISE 1.6.– *Inter-arrivals*

Inter-arrivals follow an exponential distribution with parameter λ. These arrivals arrive according to a Poisson process with parameter λ. We use $N(s;t)$ to denote its counting measure for s and t expressed in hours. This counting measure follows a Poisson distribution with parameter $\lambda(t - s)$.

The probability that no call will arrive during the first half-hour is equal to:

$$p_1 = \mathbb{P}(N(8;8{,}5) = 0) = \frac{(0{,}5\lambda)^0}{0!} e^{-0{,}5\lambda} = e^{-0{,}5\lambda}$$

The probability that no call will arrive during the second half-hour of work given that no call arrived during the first half-hour is equal to:

$$p_2 = \mathbb{P}(N(8{,}5;9) = 0 | N(8;8{,}5) = 0)$$

Given the amnesia property of the Poisson process, the number of calls arriving in any interval does not depend on the number of arrivals in any other interval. Consequently: $\mathbb{P}(N(8{,}5;9)=0|N(8;8{,}5)=0)=\mathbb{P}(N(8{,}5;9)=0)$.

$$p_2=\mathbb{P}(N(8{,}5;9)=0)=e^{-0{,}5\lambda}$$

The second call arrives during the first half-hour of work: this is equivalent to saying that the number of calls in the first half-hour of work is greater than or equal to two calls. Its probability is:

$$\begin{aligned} p_3 &= \mathbb{P}(N(8;8{,}5)\geq 2)=1-\mathbb{P}(N(8;8{,}5)=0)-\mathbb{P}(N(8;8{,}5)=1)\\ &=1-\frac{(0{,}5\lambda)^0}{0!}e^{-0{,}5\lambda}-\frac{(0{,}5\lambda)^1}{1!}e^{-0{,}5\lambda}=1-(1+0{,}5\lambda)e^{-0{,}5\lambda}\end{aligned}$$

The n-th call only arrives at instant t: it is the equivalent of saying that the number of calls at instant t is less than or equal to n. Its probability is:

$$p_4=\mathbb{P}(N(0;t)\leq n)=\sum_{k=0}^{n}\frac{(0{,}5\lambda)^k}{k!}e^{-0{,}5\lambda}$$

EXERCISE 1.7.– *ATM*

Let $N(t)$ be the counting process of the Poisson process of arrivals of customers at the ATM between instant 0 and instant t expressed in hours. We have here a Poisson process with intensity $\lambda=4$ customers per hour.

The probability that no customers will arrive in the space of one hour is:

$$p_1=\mathbb{P}(N(1)=0)=e^{-\lambda}=e^{-4}=0{,}018$$

The probability of n successive customers arriving in an interval of one minute is:

$$p_2=\mathbb{P}\left(N\left(\frac{1}{60}\right)=n\right)=\frac{\left(\frac{\lambda}{60}\right)^n}{n!}e^{-\frac{\lambda}{60}}=\frac{1}{n!\,15^n}e^{-\frac{1}{15}}=\frac{0{,}934}{n!\,15^n}$$

The average time since the last customer arrived is equal to the average inter-arrival: 15 minutes.

EXERCISE 1.8.– *Bus*

The arrival of any bus follows a Poisson process that is the superposition of two Poisson processes with respective intensities λ_1 and λ_2. The superposed process has an intensity $\lambda = \lambda_1 + \lambda_2$.

Given the amnesia property of the exponential distribution (the inter-arrival of buses), the delay time for the bus also follows an exponential distribution with parameter λ. Thus, the average delay time for the next bus is equal to $1/(\lambda_1 + \lambda_2)$.

The probability that the Line 2 bus will arrive first at stop A is equal to λ_2/λ.

Again, using the amnesia of the exponential distribution of the inter-arrivals of the Line 1 bus, and by symmetry, the average time that has passed since the last bus of this line came is equal to $1/\lambda_1$.

By the same principle, the average delay time for the next Line 1 bus is also $1/\lambda_1$.

By using these last two results, we find that, from the perspective of the user going to bus stop A at an arbitrary instant, the average duration separating two buses is equal to $1/\lambda_1 + 1/\lambda_1$. The first $1/\lambda_1$ designates the average time spent between when the bus comes by and when it arrives, and then the second $1/\lambda_1$ designates the time between its arrival and when the next bus comes by. The total time is therefore $2/\lambda_1$, which is the double of the average effective duration of the bus inter-arrivals.

EXERCISE 1.9.– *Gas station*

The gas station is made up of two parallel queues. Arrivals at the gas station follow a Poisson process with an intensity of 15 arrivals in 10 minutes, or $\lambda = 90$ arrivals per hour. The arriving car randomly and uniformly chooses the queue it will join. This is therefore a subdivision of a Poisson process (probability $p_1 = p_2 = 1/2$) and, consequently, arrivals at each pump form a Poisson process of intensity $\lambda_1 = \frac{90}{2} = 45$ arrivals per hour. We respectively use $N_1(t)$ and $N_2(t)$ to denote the counting measures of the Poisson processes on pumps 1 and 2.

The probability that no car will join the first queue at the station within the space of one hour is: $\mathbb{P}(N_1(1) = 0) = e^{-45}$.

The average number of cars arriving in the second queue within the space of one hour is: $E\big(N_2(1)\big) = 45$ cars.

The probability that more than one minute will pass between two arrivals in the second queue is: $\mathbb{P}\left(\tau_2 > \frac{1}{60}\right)$, where τ_2 indicates the inter-arrival in the second queue. τ_2 follows an exponential distribution with parameter $\lambda_2 = 45$ arrivals per hour.

$$\mathbb{P}\left(\tau_2 > \frac{1}{60}\right) = e^{-\lambda_2 t} = e^{-45/60} = e^{-0.75} = 0.47$$

EXERCISE 1.10.– *Interlacing*

We use the superposition property of Poisson process. The superposition of two processes forms a Poisson process of intensity $\lambda = \lambda_1 + \lambda_2$, of which each point comes from the first process with a probability of λ_1/λ, and from the second process with a probability of λ_2/λ. The number of arrivals X from the first process between two arrivals from the second therefore follows a geometric distribution with parameter λ_2/λ over $\mathbb{N}$:

$$\forall n \geq 0, \qquad \mathbb{P}(X = n) = \frac{\lambda_2}{\lambda}\left(\frac{\lambda_1}{\lambda}\right)^n$$

EXERCISE 1.11.– *Discretization of the Poisson process*

We use $N_B(s,t)$ to denote the counting measure of the Bernoulli process in the interval (s,t). To simplify notation, we posit that $t = s + n$, $n, \in \mathbb{N}^*$. The even $N_B(s,t) = k$ is equivalent to saying that among the n instants between s and t, we can find k instants where the process appears:

$$\mathbb{P}(N_B(s,t) = k) = \binom{n}{k} p^k (1-p)^{n-k} = \binom{t-s}{k} p^k (1-p)^{t-s-k}$$

For an intensity $p = \lambda\tau$, the counting measure of the Bernoulli process is defined by its distribution:

$$\mathbb{P}(N_B(s,t) = k) = \binom{t-s}{k} (\lambda\tau)^k (1-\lambda\tau)^{t-s-k}$$

We posit $m = 1/\tau$, and when τ tends towards zero, m tends towards infinity. Now, the binomial distribution $\mathcal{B}(m, \lambda/m) \to \mathcal{P}(\lambda)$ when m tends towards infinity.

$$\mathbb{P}(N_B(s,t) = k) = \binom{t-s}{k} \left(\frac{\lambda}{m}\right)^k \left(1 - \frac{\lambda}{m}\right)^{t-s-k} \to \frac{\left(\lambda(t-s)\right)^k}{k!} e^{-\lambda(t-s)}$$

The counting measure as a distribution tends towards the Poisson distribution with parameter $\lambda(t-s)$, the one that the counting measure of the Poisson process of intensity λ follows.

Consequently, the Bernoulli process tends towards the Poisson process for a step τ tending towards zero.

10

Chapter 2 Exercises

EXERCISE 2.1.– *Ergodic*

P is a transition matrix for a Markov chain in discrete time since the sum of the coordinates in each row equals 1. The transition graph of the corresponding chain is thus:

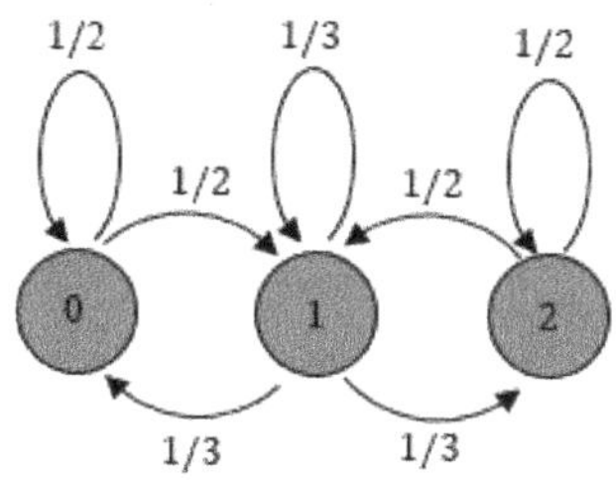

Figure 10.1. *Transition graph for Exercise 2.1*

$$\mathbf{P}^2 = \begin{pmatrix} 1/2 & 1/2 & 0 \\ 1/3 & 1/3 & 1/3 \\ 0 & 1/2 & 1/2 \end{pmatrix}^2 = \begin{pmatrix} 5/12 & 5/12 & 1/6 \\ 5/18 & 4/9 & 5/18 \\ 1/6 & 5/12 & 5/12 \end{pmatrix}$$

We found a power of **P** for which all of the coordinates are strictly positive. The conditions imposed by Proposition 2.4 are satisfied. Thus, the Markov chain is ergodic.

EXERCISE 2.2.– *Upper triangular matrix*

P is the transition matrix of a Markov chain in discrete time since the sum of the coordinates of each row equals 1.

In algebra, the product of two upper triangular matrices is an upper triangular matrix. Thus, all the powers of matrix **P** are upper triangular matrices. Consequently, all matrices $\mathbf{P}^n$ have at least one coordinate that is zero, which contradicts the condition from Proposition 2.4.

For the stationary distribution, the solution to the equation $\boldsymbol{\pi} = \boldsymbol{\pi}\mathbf{P}$ gives the following system of equations:

$$\begin{cases} \pi_0 = \pi_0 p^2 \\ \pi_1 = \pi_0 . 2p(1-p) + \pi_1 p \\ \pi_2 = \pi_0 (1-p)^2 + \pi_1 (1-p) + \pi_2 \\ \ 1 = \pi_0 + \pi_1 + \pi_2 \end{cases}$$

The solution gives us $\boldsymbol{\pi} = (\pi_0, \pi_1, \pi_2) = (0,0,1)$, so the stationary distribution exists.

We cannot say anything about the ergodicity of the chain. However, what we have found here is that the existence of the stationary distribution does not depend on the ergodicity.

EXERCISE 2.3.– *Stationary distribution*

For $\mathbf{P} = \begin{pmatrix} 1/2 & 0 & 1/2 \\ 1 & 0 & 0 \\ 1/4 & 1/4 & 1/2 \end{pmatrix}$, the associated system of equations for calculating the stationary distribution is given by:

$$\begin{cases} \pi_0 = \frac{1}{2}\pi_0 + \pi_1 + \frac{1}{4}\pi_2 \\ \pi_1 = \frac{1}{4}\pi_2 \\ \pi_2 = \frac{1}{2}\pi_0 + \frac{1}{2}\pi_2 \\ \ 1 = \pi_0 + \pi_1 + \pi_2 \end{cases}$$

The solution gives us: $\boldsymbol{\pi} = (\pi_0, \pi_1, \pi_2) = (4/9, 1/9, 4/9)$.

For $\mathbf{P} = \begin{pmatrix} p(2-p) & (1-p)^2 \\ p & 1-p \end{pmatrix}$, $0 \le p \le 1$, the system of equations for finding the stationary distribution is given by:

$$\begin{cases} \pi_0 = \pi_0 p(2-p) + \pi_1 p \\ \pi_1 = \pi_0(1-p)^2 + \pi_1(1-p) \\ 1 = \pi_0 + \pi_1 \end{cases}$$

Solving this system gives us:

$$\boldsymbol{\pi} = (\pi_0, \pi_1) = \left(\frac{p}{1-p+p^2}, \frac{(1-p)^2}{1-p+p^2}\right)$$

EXERCISE 2.4.– *Non-existence of a stationary distribution*

We have the matrix $\mathbf{P} = \begin{pmatrix} 0 & 1 \\ 1 & 0 \end{pmatrix}$. The stationary distribution is defined by the equations $\pi_0 = \pi_1$ and $\pi_0 + \pi_1 = 1$. These two equations give us $\boldsymbol{\pi} = (\pi_0, \pi_1) = (1/2, 1/2)$. All distributions $\boldsymbol{\pi}(t)$ will be equal to $(1/2, 1/2)$.

If the distribution at instant $t = 0$ is different from $\boldsymbol{\pi}(0) = (1/2, 1/2)$, then this distribution $\lim_{t\to\infty} \boldsymbol{\pi}(t)$ does not exist since the limit $(1/2, 1/2)$ will never be reached. Indeed, if $\boldsymbol{\pi}(0) = (a, b)$ for $a + b = 1, 0 < a, b < 1$, then $\boldsymbol{\pi}(2k) = (a, b)$ and $\boldsymbol{\pi}(2k+1) = (b, a)$.

EXERCISE 2.5.– *Bee*

Hypothetically, we observe that the field that the bee goes to visit does not depend on the fields that it visited previously, but only on the one it is visiting at present. We can thus model the field visited by the bee with a Markov chain in discrete time whose states at instant t are $X(t) \in \{1,2,3\}$.

Leaving field 1, the bee goes to the next field (number 2) with a probability equal to $p_{12} = 1$, and returns to the previous field (number 1) with a zero probability p_{11}. It returns to field number 3 with a zero probability p_{13} as well.

Leaving field 2, the bee goes to the next field (number 3) with a probability equal to $p_{23} = 1 - p$ and returns to the previous field (number 2) with a probability $p_{22} = p$. It returns to field number 1 with a zero probability p_{21} as well.

Leaving field 3, the bee goes to the next field (number 1) with a probability equal to $p_{31} = 1$ and returns to the previous field (number 3) with a zero probability p_{33}. It returns to field number 1 with a zero probability p_{31} as well.

The transition matrix is thus equal to:

$$\mathbf{P} = \begin{pmatrix} p_{11} & p_{12} & p_{13} \\ p_{21} & p_{22} & p_{23} \\ p_{31} & p_{32} & p_{33} \end{pmatrix} = \begin{pmatrix} 0 & 1 & 0 \\ 0 & p & 1-p \\ 1 & 0 & 0 \end{pmatrix}$$

Its transition graph is given in Figure 10.2:

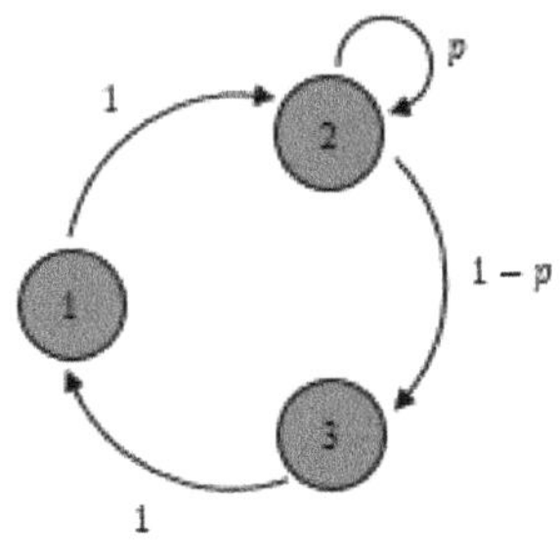

Figure 10.2. *Transition graph for Exercise 2.5*

According to the theorem of ergodicity in Proposition 2.6, the fraction of time it spends in each field is given by its stationary distribution $\boldsymbol{\pi}$. This stationary distribution is calculated using the equation $\boldsymbol{\pi} = \boldsymbol{\pi}\mathbf{P}$.

Let $\boldsymbol{\pi} = (\pi_1, \pi_2, \pi_3)$.

$$(\pi_1, \pi_2, \pi_3) = (\pi_1, \pi_2, \pi_3) \begin{pmatrix} 0 & 1 & 0 \\ 0 & p & 1-p \\ 1 & 0 & 0 \end{pmatrix}$$

$$\begin{cases} \pi_1 = \pi_3 \\ \pi_2 = \pi_1 + p\pi_2 \\ \pi_3 = (1-p)\pi_2 \\ \pi_1 + \pi_2 + \pi_3 = 1 \end{cases}$$

Solving this system of equations gives us:

$$(\pi_1, \pi_2, \pi_3) = \left(\frac{1-p}{3-2p}, \frac{1}{3-2p}, \frac{1-p}{3-2p} \right)$$

The bee spends a fraction of time $(1-p)/(3-2p)$ in field 1, a fraction of time $1/(3-2p)$ in field 2 and a fraction of time $(1-p)/(3-2p)$ in field 3.

EXERCISE 2.6.– *Traffic info*

We can describe the sequence of cars as a Markov chain over space $\mathcal{X} = \{C, V\}$, for which the transition graph is as follows:

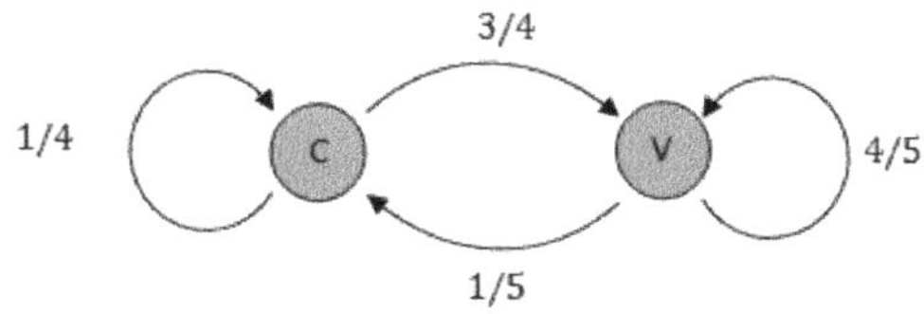

Figure 10.3. *Transition graph for Exercise 2.6*

The proportion of cars in traffic is thus given by π_V, where $\boldsymbol{\pi}$ designates the stationary distribution of this Markov chain. Here, we obtain $\pi_V = 15/19$.

EXERCISE 2.7.– *Umbrella*

X_n is the number of umbrellas that Rakoto has in any one place before beginning the n-th journey. The number of umbrellas that Rakoto has in that place at the end of the n-the journey is equal to X_{n+1}, which only depends on X_n but not on the number of umbrellas in any one place before journey n. X_n is therefore a Markov chain.

The states of this chain are: $X_n \in \{0,1,2,3\}$.

Let us suppose that, initially, Rakoto was at home (or at work, as you wish).

$p_{00} = 0$. Indeed, if no umbrella is at his home before beginning the journey, then all three umbrellas are at his office, that is, there are three on-site umbrellas before beginning the next journey. Likewise for $p_{01} = p_{02} = 0$, and we have $p_{03} = 1$.

One umbrella at home, therefore two at the office. If it rains, he takes it to his office, and before beginning the next journey, there will be three umbrellas. If it is not raining, he does not take it, there will be two umbrellas before the beginning of the next journey. Recall that the next journey is that from his office to his home and that the probability that it will rain is equal to $1/3$. Thus, $p_{13} = 1/3$ and $p_{12} = 2/3$. The other p_{1i} are zero.

Two umbrellas at home, therefore one at the office. If it is raining, he takes one to the office, and before beginning the next journey, there will be two umbrellas. If it does not rain, he will not take it, and there will be one umbrella before the beginning of the next journey. Thus, $p_{22} = 1/3$ and $p_{21} = 2/3$. The other p_{2i} are zero.

Three umbrellas at home, therefore zero at the office. If it rains, he takes one to the office, and before beginning the next journey, there will be one umbrella. If it does not rain, he will not take it, and there will be zero umbrellas before beginning the next journey. Thus, $p_{31} = 1/3$ and $p_{30} = 2/3$. The other p_{3i} are zero.

The transition matrix is therefore equal to:

$$\mathbf{P} = \begin{pmatrix} 0 & 0 & 0 & 1 \\ 0 & 0 & 2/3 & 1/3 \\ 0 & 2/3 & 1/3 & 0 \\ 2/3 & 1/3 & 0 & 0 \end{pmatrix}$$

The transition graph is presented in Figure 10.4.

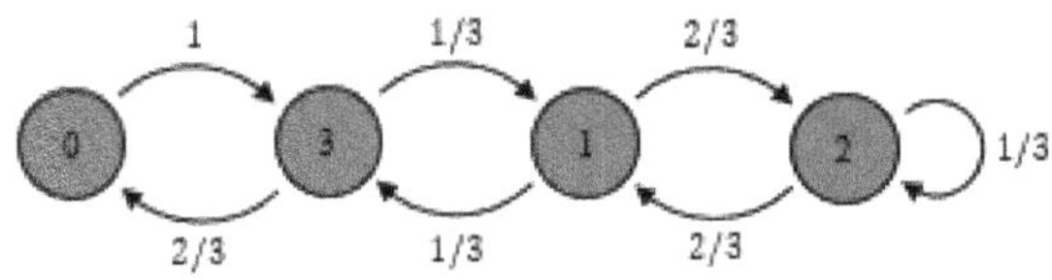

Figure 10.4. *Transition graph for Exercise 2.7*

The probability that, after a large number of journeys, Rakoto will not have an on-site umbrella when leaving is equal to the stationary probability π_0 of state 0. We will therefore determine the stationary distribution $\boldsymbol{\pi}$.

$$(\pi_0, \pi_1, \pi_2, \pi_3) = (\pi_0, \pi_1, \pi_2, \pi_3) \begin{pmatrix} 0 & 0 & 0 & 1 \\ 0 & 0 & 2/3 & 1/3 \\ 0 & 2/3 & 1/3 & 0 \\ 2/3 & 1/3 & 0 & 0 \end{pmatrix}$$

$$\begin{cases} \pi_0 = \dfrac{2\pi_3}{3} \\ \pi_1 = \dfrac{2\pi_2}{3} + \dfrac{\pi_3}{3} \\ \pi_2 = \dfrac{2\pi_1}{3} + \dfrac{\pi_2}{3} \\ \pi_3 = \pi_0 + \dfrac{\pi_1}{3} \\ \pi_0 + \pi_1 + \pi_2 + \pi_3 = 1 \end{cases} \qquad \begin{cases} 3\pi_0 = 2\pi_3 \\ 3\pi_1 = 2\pi_2 + \pi_3 \\ 3\pi_2 = 2\pi_1 + \pi_2 \\ 3\pi_3 = 3\pi_0 + \pi_1 \\ \pi_0 + \pi_1 + \pi_2 + \pi_3 = 1 \end{cases}$$

The solution to this system of equations gives us:

$$(\pi_0, \pi_1, \pi_2, \pi_3) = \left(\frac{2}{11}, \frac{3}{11}, \frac{3}{11}, \frac{3}{11}\right)$$

The probability that, after a large number of journeys, Rakoto will not have an on-site umbrella at the moment he leaves is thus equal to 2/11.

The probability that he will stupidly get wet in this case, that is, he will not have access to one of his umbrellas if it is raining when he leaves, is equal to $\frac{2}{11} \times \frac{1}{3} = 2/33$.

EXERCISE 2.8.– *Traffic*

The pedestrian goes from a lane i to lane $i+1$ with a probability of 0.2 that no car will arrive. They stay at the edge of lane i with a probability of 0.8.

The transition graph of this Markov chain is given by:

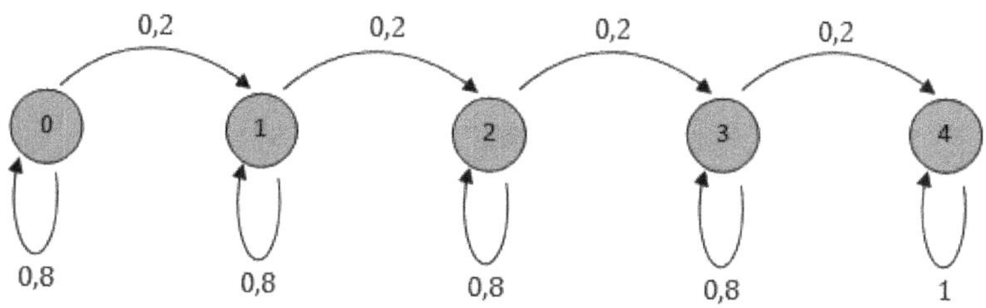

Figure 10.5. *Transition graph for Exercise 2.8*

The associated transition matrix is:

$$\mathbf{P} = \begin{pmatrix} 0.8 & 0.2 & 0 & 0 & 0 \\ 0 & 0.8 & 0.2 & 0 & 0 \\ 0 & 0 & 0.8 & 0.2 & 0 \\ 0 & 0 & 0 & 0.8 & 0.2 \\ 0 & 0 & 0 & 0 & 1 \end{pmatrix}$$

If a pedestrian can cross the four lanes after t seconds, then the state of the Markov chain is at state 4 after time t given that the initial state is state 0. The probability of crossing a lane in t seconds is therefore equal to $\pi_4(t)$ if $\boldsymbol{\pi}(0) = (1,0,0,0,0)$.

The distribution at instant t is equal to $\boldsymbol{\pi}(t) = \boldsymbol{\pi}(0)\mathbf{P}^t$, and we are seeking $\pi_4(t)$.

After calculations, the probability of crossing the highway in 4 seconds is equal to 0.0016.

The probability of crossing the highway in 6 seconds is equal to 0.017.

The probability of crossing the highway in 8 seconds is equal to 0.0563.

EXERCISE 2.9.– *Printer*

According to the hypotheses, the state of the printer at instant t only depends on its state at instant $t-1$. Thus, we can model this system as a Markov chain with three states.

The associated transition graph is given in Figure 10.6.

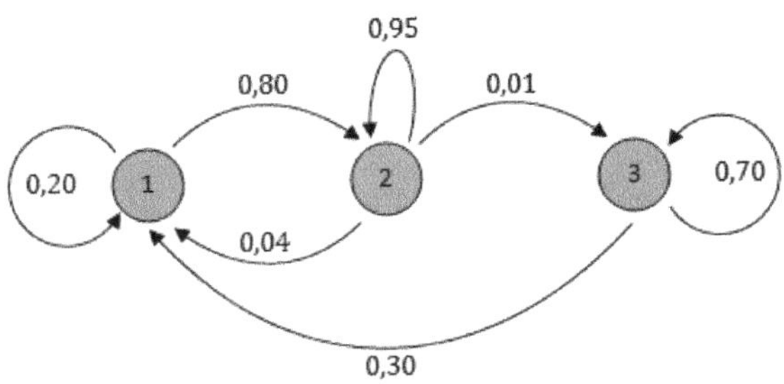

Figure 10.6. *Transition graph for Exercise 2.9.*

Its transition matrix is:

$$\mathbf{P} = \begin{pmatrix} 0.20 & 0.80 & 0 \\ 0.04 & 0.5 & 0.01 \\ 0.30 & 0 & 0.70 \end{pmatrix}$$

The balance equations for this system are:

$$\begin{aligned} \pi_1 &= 0.20\pi_1 + 0.80\pi_2 \\ \pi_2 &= 0.04\pi_1 + 0.95\pi_2 + 0.01\pi_3 \\ \pi_3 &= 0.30\pi_1 + 0.70\pi_2 \end{aligned}$$

All we need is to solve for $\mathbf{P}^2$ in order to demonstrate the ergodicity of this chain, given Proposition 2.4.

The stationary distributions are: $\pi = (1/3,\ 1/3,\ 1/3)$.

In a steady state, the rate of use of the printer is given by $\pi_2 = 1/3$.

EXERCISE 2.10.– *Fleet of buses*

Let us use $X(t)$ to denote the number of operational buses in this fleet. $X(t) \in \{0,1,\dots,N\}$. Time t is a real positive number.

X goes from state n to state $n-1$ $(n = 1,\dots,N)$ after a duration of time ε, meaning that one bus in the fleet has broken down during this time. Each vehicle breaks down independently of the others with a rate of μ and is sent to the garage, that is, within a duration of time ε, each bus has a probability equal to $\mu\varepsilon$ of breaking down. In other words, the probability that one bus from the n breaks down within a duration of time ε is equal to $n\mu\varepsilon$.

X goes from state n to state $n+1$ $(n = 0,\dots,N-1)$ after a duration of time ε, meaning that the mechanics at the garage were able to fix a bus within a duration of time that is less than ε. The time spent on their work is distributed according to an exponential distribution with parameter λ, that is, the probability that the duration of repairs will not be greater than ε is $1 - e^{-\lambda\varepsilon}$. For an infinitely small ε, we can write $1 - e^{-\lambda\varepsilon} = \lambda\varepsilon + o(\varepsilon)$.

X remains in state n $(n = 1,\dots,N-1)$ after a duration of time ε, meaning that there is no breakdown nor any repairs made during this time. Its probability is equal to $1 - n\mu\varepsilon - \lambda\varepsilon + o(\varepsilon)$. X remains in state $n = 0$ after a duration ε, meaning that there is no repair made during this time; its probability is equal to $1 - \lambda\varepsilon + o(\varepsilon)$. X remains in its state $n = N$ after a duration of time ε, meaning that there is no breakdown during this time; its probability is equal to $1 - n\mu\varepsilon$.

The transition matrix for an infinitely small duration of time ε is thus equal to, ignoring the quantities $o(\varepsilon)$:

$$\mathbf{P}(\varepsilon) = \begin{pmatrix} 1-\lambda\varepsilon & \lambda\varepsilon & 0 & 0 & 0 \\ \mu\varepsilon & 1-(\lambda+\mu)\varepsilon & \lambda\varepsilon & 0 & 0 \\ 0 & 2\mu\varepsilon & 1-(\lambda+2\mu)\varepsilon & \lambda\varepsilon & 0 \\ \vdots & \ddots & \ddots & \ddots & \ddots \\ 0 & 0 & 0 & N\mu\varepsilon & 1-N\mu\varepsilon \end{pmatrix}$$

Using equation $\mathbf{P}(\varepsilon) = \mathbf{I} + \varepsilon\mathbf{A}$, we deduce the infinitesimal stochastic generator for the chain:

$$\mathbf{A} = \begin{pmatrix} -\lambda & \lambda & 0 & 0 & 0 \\ \mu & -(\lambda+\mu) & \lambda & 0 & 0 \\ 0 & 2\mu & -(\lambda+2\mu) & \lambda & 0 \\ \vdots & \ddots & \ddots & \ddots & \ddots \\ 0 & 0 & 0 & N\mu & -N\mu \end{pmatrix}$$

The associated transition graph is presented in Figure 10.7.

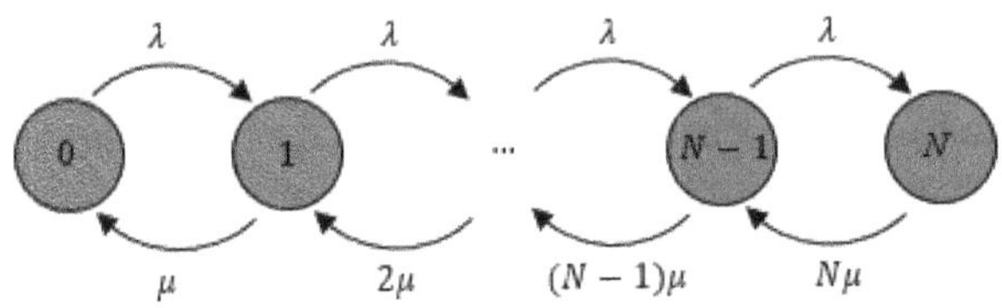

Figure 10.7. *Transition graph for Exercise 2.10*

Its stationary distribution is obtained by solving the matrix equation $\boldsymbol{\pi}\mathbf{A} = \mathbf{0}$.

Referring to equation [2.36], we find that:

$$\pi_k = \frac{\frac{1}{k!}\left(\frac{\lambda}{\mu}\right)^k}{1 + \frac{\lambda}{\mu} + \frac{1}{2!}\left(\frac{\lambda}{\mu}\right)^2 + \cdots + \frac{1}{N!}\left(\frac{\lambda}{\mu}\right)^N}$$

The average number of buses in service is equal to $\left(\frac{\lambda}{\mu}\right)(1 - \pi_N)$.

Indeed:

$$E(n) = \sum_{k=0}^{N} k\pi_k = \sum_{k=1}^{N} k\pi_k = \sum_{k=1}^{N} \frac{\lambda}{\mu}\pi_{k-1}$$

since:

$$k\pi_k = \frac{\frac{k}{k!}\left(\frac{\lambda}{\mu}\right)^k}{1 + \frac{\lambda}{\mu} + \frac{1}{2!}\left(\frac{\lambda}{\mu}\right)^2 + \cdots + \frac{1}{N!}\left(\frac{\lambda}{\mu}\right)^N} = \frac{\frac{1}{(k-1)!}\left(\frac{\lambda}{\mu}\right)^k}{1 + \frac{\lambda}{\mu} + \frac{1}{2!}\left(\frac{\lambda}{\mu}\right)^2 + \cdots + \frac{1}{N!}\left(\frac{\lambda}{\mu}\right)^N}$$

$$E(n) = \frac{\lambda}{\mu}\sum_{k=1}^{N}\pi_{k-1} = \frac{\lambda}{\mu}\sum_{i=0}^{N-1}\pi_i = \left(\frac{\lambda}{\mu}\right)(1-\pi_N)$$

EXERCISE 2.11.– *Travel*

Let $X(t)$ be the Markov chain describing the businessman's itinerary: $X(t) \in \{T, M, A\}$, where T designates Antananarivo, M Mahajanga and A Antsiranana. We will ignore the duration of travel in what follows.

Assume that at instant 0, he is in Antananarivo. The event "$X(\varepsilon) = T|X(0) = T$" is equivalent to "the time he spends in Antananarivo is greater than ε". Thus: $\mathbb{P}(X(\varepsilon) = T|X(0) = T) = e^{-4\varepsilon}$. Indeed, the time he spends in one city is exponential with an average of $1/4$. For an infinitely small ε, we can approximate it using its expansion with a limit of approximately 0: $e^{-4\varepsilon} = 1 - 4\varepsilon + o(\varepsilon)$.

The event "$X(\varepsilon) = M|X(0) = T$" is equivalent to "the time he spends in Antananarivo is less than ε, and he went to Mahajanga". Thus, $\mathbb{P}(X(\varepsilon) = M|X(0) = T) = (1 - e^{-4\varepsilon}) \times \frac{1}{2} = 2\varepsilon + o(\varepsilon)$.

The event "$X(\varepsilon) = A|X(0) = T$" is equivalent to "the time he spends in Antananarivo is less than ε, and he went to Antsiranana". Thus, $\mathbb{P}(X(\varepsilon) = A|X(0) = T) = (1 - e^{-4\varepsilon}) \times \frac{1}{2} = 2\varepsilon + o(\varepsilon)$.

By proceeding in the same manner, we find that:

$$\mathbb{P}(X(\varepsilon) = T|X(0) = M) = (1 - e^{-4\varepsilon}) \times \frac{3}{4} = 3\varepsilon + o(\varepsilon)$$

$$\mathbb{P}(X(\varepsilon) = M|X(0) = M) = e^{-4\varepsilon} = 1 - 4\varepsilon + o(\varepsilon)$$

$$\mathbb{P}(X(\varepsilon) = A|X(0) = M) = (1 - e^{-4\varepsilon}) \times \frac{1}{4} = \varepsilon + o(\varepsilon)$$

$$\mathbb{P}(X(\varepsilon) = T|X(0) = A) = (1 - e^{-5\varepsilon}) \times 1 = 5\varepsilon + o(\varepsilon)$$

$$\mathbb{P}(X(\varepsilon) = M|X(0) = A) = (1 - e^{-5\varepsilon}) \times 0 = 0 + o(\varepsilon)$$

$$\mathbb{P}(X(\varepsilon) = A|X(0) = A) = e^{-5\varepsilon} = 1 - 5\varepsilon + o(\varepsilon)$$

We can deduce the transition matrix for 0 to instant ε:

$$\mathbf{P}(\varepsilon) = \mathbf{I} + \varepsilon\mathbf{A} + o(\varepsilon) = \begin{pmatrix} 1-4\varepsilon & 2\varepsilon & 2\varepsilon \\ 3\varepsilon & 1-4\varepsilon & \varepsilon \\ 5\varepsilon & 0 & 1-5\varepsilon \end{pmatrix} + o(\varepsilon)$$

and we find the infinitesimal stochastic generator:

$$\mathbf{A} = \begin{pmatrix} -4 & 2 & 2 \\ 3 & -4 & 1 \\ 5 & 0 & -5 \end{pmatrix}$$

The transition graph of the system is thus presented in Figure 10.8.

The fraction of time that he travels from each city can be obtained from the solution to the state equation $\boldsymbol{\pi}\mathbf{A} = \mathbf{0}$, taking the theorem on ergodicity into account.

$$(\pi_T, \pi_M, \pi_A) \begin{pmatrix} -4 & 2 & 2 \\ 3 & -4 & 1 \\ 5 & 0 & -5 \end{pmatrix} = (0,0,0)$$

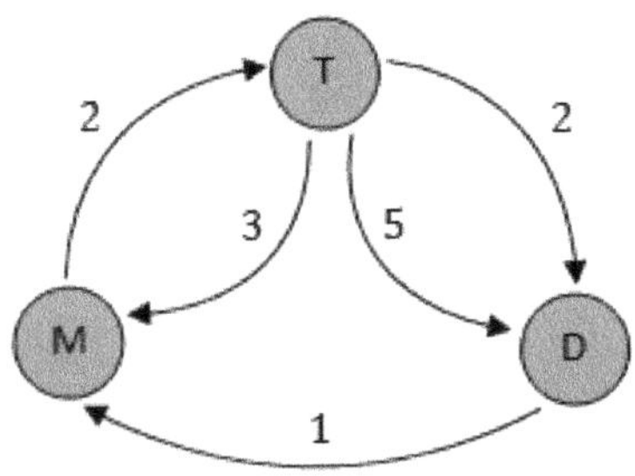

Figure 10.8. *Transition graph for Exercise 2.11*

By solving this matrix equation, we find that $\boldsymbol{\pi} = (1/2, 1/4, 1/4)$. Thus, he travels 50% of the time to Antananarivo and 25% of the time to Mahajanga and Antsiranana.

The businessman makes an average of 12 trips from Antananarivo to Mahajanga each year.

EXERCISE 2.12.– *Stock*

The transition graph for this small store is given by:

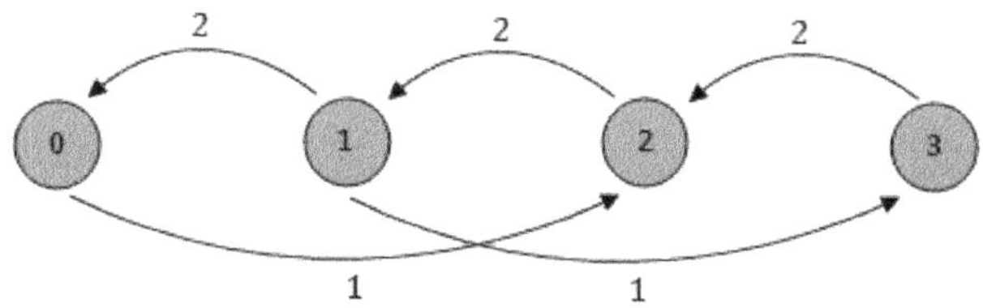

Figure 10.9. *Transition graph for Exercise 2.12*

Proceeding with the same method used in Exercise 2.11, we find the associated generator:

$$\mathbf{A} = \begin{pmatrix} -1 & 0 & 1 & 0 \\ 2 & -3 & 0 & 1 \\ 0 & 2 & -2 & 0 \\ 0 & 0 & 2 & -2 \end{pmatrix}$$

The stationary distribution of this system is equal to $\boldsymbol{\pi} = (0.4; 0.2;\ 0.3;\ 0.1)$.

The rate of sales is equal to 1.2 computers.

EXERCISE 2.13.– *Call center*

Let $X(t)$ be the Markov chain describing the number of busy operators: $X(t) \in \{0,1,\dots,s\}$.

Calls arrive according to a Poisson process with a rate of λ; thus, the probability of transition from state n to $n+1$ $(n = 0,\dots,s-1)$ follows an exponential distribution with parameter λ. For an infinitely small duration of time ε, $\mathbb{P}(X(\varepsilon) = n+1|X(0) = n) = 1 - e^{-\lambda\varepsilon} = \lambda\varepsilon + o(\varepsilon)$.

The duration of a call follows an exponential distribution with an average of $1/\mu$. The probability of a transition from state n to $n-1$ $(n = 1,\dots,s)$ is therefore equal to the probability that a call from among the n calls in progress will cease: $\mathbb{P}(X(\varepsilon) = n+1|X(0) = n) = n(1 - e^{-\mu\varepsilon}) = n\mu\varepsilon + o(\varepsilon)$.

The probability of remaining in state n for a duration of time ε is equal to that of having no new calls and having no calls cease during this duration of time: $\mathbb{P}(X(\varepsilon) = n+1|X(0) = n) = 1 - (\lambda\varepsilon + n\mu\varepsilon) + o(\varepsilon)$.

Referring to Exercise 2.10, the stationary distribution has the coordinate:

$$\pi_k = \frac{\frac{1}{k!}\left(\frac{\lambda}{\mu}\right)^k}{1 + \frac{\lambda}{\mu} + \frac{1}{2!}\left(\frac{\lambda}{\mu}\right)^2 + \cdots + \frac{1}{s!}\left(\frac{\lambda}{\mu}\right)^s}$$

The probability that a call will be rejected is equal to that of finding all s operators busy:

$$\pi_s = \frac{\frac{1}{s!}\left(\frac{\lambda}{\mu}\right)^s}{1 + \frac{\lambda}{\mu} + \frac{1}{2!}\left(\frac{\lambda}{\mu}\right)^2 + \cdots + \frac{1}{s!}\left(\frac{\lambda}{\mu}\right)^s}$$

11

Chapter 3 Exercises

EXERCISE 3.1.– *Length of a queue*

According to Little's law from Proposition 3.2:

$$\begin{cases} E(N_s) = \lambda_e E(T_s) \\ E(N_f) = \lambda_e E(T_f) \end{cases}$$

Now, $E(T_s) = E(T_f) + 1/\mu$ since we assume that the duration of service for a customer (on average $1/\mu$) is independent of the duration of time spent waiting in the queue.

Therefore, $E(N_s) = \lambda_e(E(T_f) + 1/\mu) = \lambda_e E(T_f) + \lambda_e/\mu = E(N_f) + \lambda_e/\mu$. We deduce that $E(N_f) = E(N_s) - \lambda_e/\mu$.

EXERCISE 3.2.– *M/M/1 queue*

We have an M/M/1 queuing system. A customer arrives on average every 10 minutes: $\lambda = 1/10$, and the average duration of service is 7 minutes: $\mu = 1/7$. The offered traffic is $\rho = \frac{\lambda}{\mu} = \frac{7}{10} = 0.7$. We use π_k to denote the coordinate of the stationary distribution of the system: $\pi_k = (1-\rho)\rho^k$.

The probability p_1 that at least two customers will wait to be served is equal to the probability that the system will contain at least three customers (one served and at least two waiting):

$$p_1 = \sum_{k=3}^{\infty} \pi_k = \sum_{k=3}^{\infty} (1-\rho)\rho^k = (1-\rho)\frac{\rho^3}{1-\rho} = \rho^3 = 0.7^3 = 0.343$$

The probability p_2 that an arriving customer will have to wait before being served is equal to the probability that the system is busy serving a customer:

$$p_2 = \pi_1 = (1-\rho)\rho = 0.3 \times 0.7 = 0.21$$

The probability p_3 that an arriving customer will find a queue in front of them with n people is equal to the probability that the system will contain $n+1$ customers (one served and n waiting):

$$p_3 = \pi_{n+1} = (1-\rho)\rho^{n+1} = 0.21 \times 0.7^n$$

EXERCISE 3.3.– *File transfer*

The file is transferred as blocks of files of 100,000 characters of 8 bits, or 800 kbits. The line used for the transfer has a capacity of 512 kbits per second, so the average duration of $1/\mu$ needed to transfer a block is equal to $\frac{800}{512} = 1.56$ seconds.

The line delivers Poissonian traffic with a load limited to 60%, or $\rho = 0.6$. The rate of arrival is equal to $\lambda = \rho\mu = 0.6 \times \frac{1}{1.56} = 0.38$ blocks per second.

Here, the line can be modeled by an M/M/1 queue.

The response time of the line is equal to the holding time of a block in a line. This is equal to 3.8 seconds.

$$E(T_s) = \frac{1}{\mu - \lambda} = \frac{1}{\frac{1}{1.56} - 0.38} = 3.8$$

The average delay time of the line is equal to 2.24 seconds.

$$E(T_q) = E(T_s) - \frac{1}{\mu} = 3.8 - 1.56 = 2.24$$

EXERCISE 3.4.– *Gas station*

The gas station can be modeled by an M/M/1 queue, such that $\lambda = \frac{10}{60} = \frac{1}{6}$ and $\mu = \frac{1}{2}$. $\rho = \frac{\lambda}{\mu} = \frac{1}{3}$. The unit of time is minutes.

The stationary distribution of the number of cars in the station is:

$$\pi_k = (1-\rho)\rho^k = 2/3^{k+1}$$

The average delay time before being served is 1 minute. Indeed, we have:

$$E(T_q) = E(T_s) - \frac{1}{\mu} = \frac{1}{\mu - \lambda} - \frac{1}{\mu} = \frac{\lambda}{\mu(\mu - \lambda)} = \frac{\frac{1}{3}}{\frac{1}{2} - \frac{1}{6}} = 1$$

The total holding time is 3 minutes. We have:

$$E(T_s) = \frac{1}{\mu - \lambda} = \frac{1}{\frac{1}{2} - \frac{1}{6}} = 3$$

The proportion of cars that must wait before being able to fill their tanks is 33%. This is equal to the probability that one car will wait:

$$1 - \pi_0 = 1 - 2/3 = 0.33$$

The proportion of cars that must spend more than 2 minutes in the station is 51.3%. Indeed, this is equal to the probability that a car will remain for more than 2 minutes. Considering that the holding time r of this car follows an exponential distribution with parameter $\mu - \lambda$, this proportion is equal to:

$$\mathbb{P}(T_s > 2) = e^{-(\mu-\lambda)2} = e^{-2/3} = 0.513$$

We now assume that any driver who encounters two cars in the station will immediately leave. This time, the station can be modeled by an M/M/1/2 queue.

The model of the M/M/1/2 delay system with a birth and death process gives us the following hypotheses and results. The growth rates of the process are: $\lambda_0 = \lambda_1 = \lambda$, $\lambda_i = 0$ for $i \geq 2$. The death rates are: $\mu_1 = \mu_2 = \mu$.

Thus:

$$\pi_0 = \frac{1}{1 + \frac{\lambda}{\mu} + \left(\frac{\lambda}{\mu}\right)^2} = \frac{1}{1 + \rho + \rho^2},$$

$$\pi_1 = \frac{\frac{\lambda}{\mu}}{1 + \frac{\lambda}{\mu} + \left(\frac{\lambda}{\mu}\right)^2} = \frac{\rho}{1 + \rho + \rho^2},$$

$$\pi_2 = \frac{\left(\frac{\lambda}{\mu}\right)^2}{1 + \frac{\lambda}{\mu} + \left(\frac{\lambda}{\mu}\right)^2} = \frac{\rho^2}{1 + \rho + \rho^2}$$

The probability that a car will leave without filling its tank is equal to the probability that two cars will already be in the system, that is, $\pi_2 = \frac{1}{13} = 0.077$.

Using Little's law from Proposition 3.2, $E(T_s) = E(N_s)/\lambda$, with:

$$E(N_s) = \sum_{k=0}^{2} k\pi_k = \frac{\rho + 2\rho^2}{1 + \rho + \rho^2}$$

The average holding time is equal to $E(T_s) = \frac{1}{\lambda}.\frac{\rho+2\rho^2}{1+\rho+\rho^2}$, and the average delay time is equal to $E(T_q) = E(T_s) - \frac{1}{\mu} = \frac{1}{\lambda}.\frac{\rho+2\rho^2}{1+\rho+\rho^2} - \frac{1}{\mu}$. The numerical calculations give us $E(T_s) = \frac{30}{13} = 2.3$ minutes and $E(T_q) = 1.8$ minutes.

EXERCISE 3.5.– *At the cybercafé*

The cybercafé can be modeled by an M/M/2/2 queuing system. The arrival rate is $\lambda = 3$ customers per hour, and the average duration of service is equal to $1/\mu = 0.5$ hours, or $\mu = 2$. We thus have $\rho = \frac{\lambda}{\mu} = \frac{3}{2} = 1.5$.

For $k = 0, 1, 2$, its stationary distribution is equal to:

$$\pi_k = \frac{\frac{\rho^k}{k!}}{1 + \rho + \frac{\rho^2}{2!}}$$

The calculations give us $\boldsymbol{\pi} = (0.28; 0.41; 0.31)$.

The probability that an arriving customer will not be served is equal to the probability that the two service nodes are busy, or $\pi_2 = \frac{\frac{\rho^2}{2!}}{1+\rho+\frac{\rho^2}{2!}} = 0.31$.

For the case of a mini cybercafé with a waiting room containing two chairs, we have an M/M/2/4 delay system. The parameters of the birth and death process modeling this delay system are:

– growth rate: $\lambda_k = \lambda = 3$ customers per hour, for $k = 0, 1, 2, 3$;

– death rate: $\mu_1 = \mu = 2$, $\mu_2 = \mu_3 = \mu_4 = 2\mu = 4$.

The stationary distribution is defined by:

$$\pi_k = \frac{\frac{1}{2^{k-1}}\left(\frac{\lambda}{\mu}\right)^k}{1+\left(\frac{\lambda}{\mu}\right)+\frac{1}{2}\left(\frac{\lambda}{\mu}\right)^2+\frac{1}{4}\left(\frac{\lambda}{\mu}\right)^3+\frac{1}{8}\left(\frac{\lambda}{\mu}\right)^4}, \quad k = 1,2,3,4$$

$$\pi_0 = \frac{1}{1+\left(\frac{\lambda}{\mu}\right)+\frac{1}{2}\left(\frac{\lambda}{\mu}\right)^2+\frac{1}{4}\left(\frac{\lambda}{\mu}\right)^3+\frac{1}{8}\left(\frac{\lambda}{\mu}\right)^4}$$

The calculations give us $\boldsymbol{\pi} = (0.20; 0.29; 0.22; 0.17; 0.12)$.

The probability that a customer will have to wait is equal to $\pi_3 + \pi_4 = 0.29$.

The average delay time can be obtained from Little's law: $E(T_q) = \frac{E(N_S)}{\lambda} - \frac{1}{\mu}$, with $E(N_s) = \sum k\pi_k = 1.73$. Thus, $E(T_q) = 0.075$ hour, or 4.5 minutes.

EXERCISE 3.6.– *Comparison of queues (1)*

We have an M(λ)/M(μ)/2 queue and an M(λ)/M(2μ)/1 queue.

The average numbers of customers in these systems are:

$$E(N_s)_{M(\lambda)/M(\mu)/2} = \frac{4\lambda\mu}{4\mu^2-\lambda^2}, \quad E(N_s)_{M(\lambda)/M(2\mu)/1} = \frac{\lambda}{2\mu-\lambda}$$

The average delay times are:

$$E(T_q)_{M(\lambda)/M(\mu)/2} = \frac{\lambda^2}{\mu(4\mu^2-\lambda^2)}, \quad E(T_q)_{M(\lambda)/M(2\mu)/1} = \frac{\lambda}{2\mu(2\mu-\lambda)}$$

The average holding times are:

$$E(T_s)_{M(\lambda)/M(\mu)/2} = \frac{4\mu}{4\mu^2-\lambda^2}, \quad E(T_s)_{M(\lambda)/M(2\mu)/1} = \frac{1}{2\mu-\lambda}$$

For the comparisons, we calculate the differences in scale between the two types of queues.

$$\frac{4\lambda\mu}{4\mu^2-\lambda^2} - \frac{\lambda}{2\mu-\lambda} = \frac{\lambda}{2\mu+\lambda} > 0, \quad E(N_s)_{M(\lambda)/M/2} > E(N_s)_{M(\lambda)/M(2\mu)/1}$$

$$\frac{\lambda^2}{\mu(4\mu^2-\lambda^2)}-\frac{\lambda}{2\mu(2\mu-\lambda)}=-\frac{\lambda}{2\mu(2\mu+\lambda)}<0,$$

$$E(T_f)_{M(\lambda)/M/2}<E(T_f)_{M(\lambda)/M(2\mu)/1}$$

$$\frac{4\mu}{4\mu^2-\lambda^2}-\frac{1}{2\mu-\lambda}=\frac{1}{2\mu+\lambda}>0,\quad E(T_s)_{M(\lambda)/M/2}>E(T_s)_{M(\lambda)/M(2\mu)/1}$$

The delay time for the M(λ)/M(2μ)/1 queue is longer than that of the M(λ)/M(μ)/2 queue; the total holding time is shorter however. The length of the M(λ)/M(2μ)/1 queue is shorter than that of the M(λ)/M(μ)/2 queue. We can say that it is preferable to replace the two servers by one that is twice as efficient.

From the two lengths, we can say that the M(λ)/M(μ)/2 queue has a greater probability of being busy than the M(λ)/M(2μ)/1 queue. This conclusion can also be verified from the calculation of the two stationary distributions.

EXERCISE 3.7.– *Comparison of queues (2)*

We have two M($\lambda/2$)/M(μ)/1 queues and one M(λ)/M(μ)/2 queue.

The average numbers of customers in these systems are:

$$E(N_s)_{M(\lambda/2)/M(\mu)/1}=\frac{\lambda}{2\mu-\lambda},\qquad for\ each\ queue$$

$$E(N_s)_{2\times M(\lambda/2)/M(\mu)/1}=\frac{2\lambda}{2\mu-\lambda},\qquad for\ both\ 2\ queues\ together$$

$$E(N_s)_{M(\lambda)/M(\mu)/2}=\frac{4\lambda\mu}{4\mu^2-\lambda^2}$$

The average delay times are:

$$E(T_q)_{2\times M(\lambda/2)/M(\mu)/1}=\frac{\lambda}{\mu(2\mu-\lambda)},\quad E(T_q)_{M(\lambda)/M(\mu)/2}=\frac{\lambda^2}{\mu(4\mu^2-\lambda^2)}$$

The average holding times are:

$$E(T_s)_{2\times M(\lambda/2)/M(\mu)/1}=\frac{2}{2\mu-\lambda},\quad E(T_s)_{M(\lambda)/M(\mu)/2}=\frac{4\mu}{4\mu^2-\lambda^2}$$

For the comparisons, we can also calculate the differences in scale between these two types of queues.

$$\frac{4\lambda\mu}{4\mu^2-\lambda^2}-\frac{\lambda}{2\mu-\lambda}=\frac{\lambda}{2\mu+\lambda}>0, \qquad E(N_s)_{M(\lambda)/M(\mu)/2}>E(N_s)_{M(\lambda/2)/M(\mu)/1}$$

$$\frac{4\lambda\mu}{4\mu^2-\lambda^2}-\frac{2\lambda}{2\mu-\lambda}=-\frac{2\lambda^2}{4\mu^2-\lambda^2}<0$$

$$E(N_s)_{M(\lambda)/M(\mu)/2}<E(N_s)_{2\times M(\lambda/2)/M(\mu)/1}$$

$$\frac{\lambda^2}{\mu(4\mu^2-\lambda^2)}-\frac{\lambda}{\mu(2\mu-\lambda)}=-\frac{2\lambda}{4\mu^2-\lambda^2}<0$$

$$E(T_q)_{M(\lambda)/M(\mu)/2}<E(T_q)_{2\times M(\lambda/2)/M(\mu)/1}$$

$$\frac{4\mu}{4\mu^2-\lambda^2}-\frac{2}{2\mu-\lambda}=-\frac{2\lambda}{4\mu^2-\lambda^2}<0$$

$$E(T_s)_{M(\lambda)/M(\mu)/2}<E(T_s)_{2\times M(\lambda/2)/M(\mu)/1}$$

In both cases, the number of servers is the same. We are therefore comparing two distributions of arrivals here. The delay times and the holding times for the $M(\lambda)/M(\mu)/2$ queue are shorter than those of the divided $M(\lambda/2)/M(\mu)/1$ queue. Likewise, the length of the $M(\lambda)/M(\mu)/2$ queue is shorter than the total length of both $M(\lambda/2)/M(\mu)/1$ queues. Consequently, from the point of view of customers, the single queue is more advantageous than the divided queue. Nevertheless, the individual length of each $M(\lambda/2)/M(\mu)/1$ queue is shorter than that of the $M(\lambda)/M(\mu)/2$ queue. In certain infrastructures, it is preferable to instead have several queues to reduce the length of each parallel queue.

From these lengths, we can conclude that the $M(\lambda)/M(\mu)/2$ queue has a greater probability of being busy than the two $M(\lambda/2)/M(\mu)/1$ queues. We can also demonstrate this with the calculation of stationary distributions.

EXERCISE 3.8.– *Telephone booths*

The average number of customers and the average holding time of customers in an M/M/*s* delay system are as follows.

For $s = 1$ service nodes:

$$\frac{1}{\pi_0} = 1 + \frac{\rho}{1-\rho} = \frac{1}{1-\rho}$$

$$E(N_s) = \rho + \pi_0 \frac{\rho^2}{(1-\rho)^2} = \frac{\rho}{1-\rho}$$

$$E(T_s) = \frac{\pi_0 \rho}{1-\rho} \frac{1}{\mu - \lambda} = \frac{\rho}{\mu(1-\rho)}$$

For $s = 2$ service nodes:

$$\frac{1}{\pi_0} = 1 + \rho + \frac{\rho^2}{2\left(1-\frac{\rho}{2}\right)} = \frac{2+\rho}{2-\rho}$$

$$E(N_s) = \rho + \pi_0 \frac{\rho^3}{(2-\rho)^2} = \frac{4\rho}{4-\rho^2}$$

$$E(T_s) = \frac{\pi_0 \rho^2}{2\left(1-\frac{\rho}{2}\right)} \frac{1}{2\mu - \lambda} = \frac{\rho^2}{\mu(4-\rho^2)}$$

For $s = 3$ service nodes:

$$\frac{1}{\pi_0} = 1 + \rho + \frac{\rho^2}{2} + \frac{\rho^3}{6\left(1-\frac{\rho}{3}\right)} = \frac{6+4\rho+\rho^2}{2(3-\rho)}$$

$$E(N_s) = \rho + \pi_0 \frac{\rho^4}{2(3-\rho)^2} = \frac{\rho(18+6\rho-\rho^2)}{(3-\rho)(6+4\rho+\rho^2)}$$

$$E(T_s) = \frac{\pi_0 \rho^3}{6\left(1-\frac{\rho}{3}\right)} \frac{1}{3\mu - \lambda} = \frac{\rho^3}{\mu(3-\rho)(6+4\rho+\rho^2)}$$

Now, we have a telephone booth with an arrival distribution following a Poisson process with a rate of 6 people/hour: $\lambda = 6/60 = 0.1$ people/minute. The time spent by an individual in the booth follows an exponential distribution with an average of 5 minutes: $\mu = 1/5 = 0.2$. Thus, $\rho = \lambda/\mu = 0.5$.

The average holding time for each individual is: $E(T_s)$ for $s = 1$, or $E(T_s) = \frac{\rho}{\mu(1-\rho)} = 10$ minutes.

The average number of people in the system is: $E(N_s)$ for $s = 1$, or $E(N_s) = \frac{\rho}{1-\rho} = 1$ person.

For the probability that the total time spent in the system will be at least 10 minutes, we have an M/M/1 queue, so the holding time follows an exponential distribution with parameter $\mu - \lambda = 0.2 - 0.1 = 0.1$. This probability is therefore equal to $e^{-0.1\times 10} = 0.37$.

Now, the arrival rate rises to 20 people/hour, or $\lambda = 20/60 = 1/3$ people/minute. The time spent by an individual remains 5 minutes on average, so $\mu = 1/5 = 0.2$. Thus, $\rho = \lambda/\mu = 5/3$. We have $\rho > 1$.

If we have s telephone booths:

– $s = 1$, $\rho > 1$, we will never reach a steady state, the length of the queue will become infinite and $E(N_s) = E(T_s) = \infty$;

– $s = 2$, $\rho < 2$, there is a steady state. After calculations, we find $E(N_s) = \frac{60}{11} = 5.45$ people and $E(T_s) = \frac{5\times 25}{11} = 11.36$ minutes;

– $s = 3$, $\rho < 3$, there is a steady state. After calculations, we find $E(N_s) = \frac{1135}{556} = 2.04$ people and $E(T_s) = \frac{5\times 125}{556} = 1.12$ minutes.

There must therefore be at least three telephone booths to have an average number of people that is less than three and a holding time that is less than 10 minutes.

EXERCISE 3.9.– *Telephone switch*

The telephone switch can be modeled by an M/M/s/s delay system. Its stationary distribution is therefore equal to:

$$\pi_k = \frac{\frac{\rho^k}{k!}}{1 + \rho + \frac{\rho^2}{2!} + \cdots + \frac{\rho^s}{s!}} \qquad (0 \le k \le s)$$

where $\rho = \lambda/\mu$.

The probability that a call will be rejected is equal to the probability that all s lines are busy, or s customers are already in the system, that is, the probability is equal to π_s:

$$\pi_s = \frac{\frac{\rho^s}{s!}}{1 + \rho + \frac{\rho^2}{2!} + \cdots + \frac{\rho^s}{s!}}$$

EXERCISE 3.10.– *Queues in tandem*

We examine the case of two consecutive customers: the $(n-1)$-th customer and the n-th customer in the system. The n-th customer begins their first service of a duration $D_n^{(1)}$ after the departure of the $(n-1)$-th customer getting this service, that is:

– immediately after the $(n-1)$-th customer leaves for the second service if they are present in the system at that instant;

– or during the second service of the $(n-1)$-th customer, or even later, if they have not yet entered the system at the instant where the $(n-1)$-th customer leaves the first service.

The n-th customer remains stuck at the first service if, at the end of it, the $(n-1)$-th customer is still at the second service, and the entire system remains blocked.

We use $D_{n-1}^{(2)*}$ to denote the service time remaining for the second service being provided to the $(n-1)$-th customer, starting at the moment when the n-th customer enters the first service. The system is thus blocked if, and only if, $D_{n-1}^{(2)*} > D_n^{(1)}$.

The probability of a blockage is therefore equal to:

$$p_B = \mathbb{P}\left(D_{n-1}^{(2)*} > D_n^{(1)}\right)$$

Using the amnesia property of the exponential distribution, we can say that the random variable $D_{n-1}^{(2)*}$ follows the same distribution as $D_{n-1}^{(2)}$, which is the exponential distribution with parameter μ_2.

Therefore, we have:

$$
\begin{aligned}
p_B &= \mathbb{P}\left(D_{n-1}^{(2)*} > D_n^{(1)}\right) = \int_0^{\infty} \mathbb{P}\left(D_{n-1}^{(2)*} > t\right) f_{D_n^{(1)}}(t)dt \\
&= \int_0^{\infty} e^{-\mu_2 t}.\mu_1 e^{-\mu_1 t} dt = \int_0^{\infty} \mu_1 e^{-(\mu_1+\mu_2)t} dt \\
&= \frac{\mu_1}{\mu_1 + \mu_2}
\end{aligned}
$$

This probability of a blockage is proportional to μ_1, so for it to be minimal, μ_1 must be smaller than μ_2. Now μ_i designates the parameter of the exponential distribution, which is the inverse of the average duration of service; so the inverse of the smaller average duration of service will be placed first. That is, the slower service will need to be placed in the first position.

12

Chapter 4 Exercise

One of the difficulties of this exercise compared to simple queues is that the state space is not $\mathbb{N}$ but $\mathbb{N}^N$, and the indexes are therefore vectors of size N.

To keep notations easy to read, we take $\mathbf{n}$ to represent the states (it will be a vector) and n_p for the p-th coordinate. To avoid confusing these notations with those of the distributions, if $\boldsymbol{\pi}$ is a distribution for the whole system (thus for the N-tuple), $\boldsymbol{\pi}_p$ represents the marginal distribution of the p-th queue and the probabilities of being in a given state will be denoted $\boldsymbol{\pi}(\mathbf{n})$ and $\boldsymbol{\pi}_p(n_p)$.

To determine the infinitesimal stochastic generator, we must decompose the types of transitions that are possible for a given queue p:

– exogenous arrivals: $\bar{\lambda}_p$;

– exiting the network: $\mu_p(n_p).r_{p,N+1}$;

– routing from queue p to queue $q \neq p$:$\mu_p(n_p).r_{p,q}$.

From this, we can deduce the non-zero terms of the infinitesimal generator:

$$q_{\mathbf{n},\mathbf{n}+\mathbf{e}_p} = \bar{\lambda}_p q_{\mathbf{n}+\mathbf{e}_p,\mathbf{n}} = \mu_p(n_p).r_{p,N+1} q_{\mathbf{n}+\mathbf{e}_p,\mathbf{n}+\mathbf{e}_q} = \mu_p(n_p).r_{p,q}$$

Let us denote λ_p the expectation of entering queue p, which combines exogenous arrivals and routing from other queues. Since the queues are in equilibrium, everything that enters leaves, so the expectation of an exit from queue p is also λ_p. A fraction $r_{p,q}$ of this quantity is sent to queue q. Adding the sum of the different types of arrivals, we obtain the following equations:

$$\forall p \in [1, N], \qquad \lambda_p = \bar{\lambda}_p + \sum_{q \neq p} \lambda_q r_{q,p}$$

This is a linear system of unknowns λ_p that we call traffic equations.

The property of "being without a capture" is formally translated as follows: for every p, there exists a sequence of coordinates $p_1, p_2, \dots, p_N$of $[1, N]$ such that $r_{p,p_1} r_{p_1,p_2} \cdots r_{p_{n-1},p_n} r_{p_n,N+1} > 0$. If we use $\mathbf{R} = \left(r_{p,q}\right)_{1 \le p,q \le N}$to denote the matrix for routing between queues, we are seeking a solution to the system $\boldsymbol{\lambda} = \bar{\boldsymbol{\lambda}} + \boldsymbol{\lambda}\mathbf{R}$ that allows for a solution if and only if the series $\mathbf{K} = \mathbf{I} + \mathbf{R} + \mathbf{R}^2 + \cdots$ converges (and in this case it equals $(\mathbf{I} - \mathbf{R})^{-1}$). The unique positive and finite solution is thus given by $\boldsymbol{\lambda} = \bar{\boldsymbol{\lambda}}\mathbf{K}$. Let us therefore show that $\mathbf{K}$ is finite. The matrix for transitions between queues (counting exits from the system) is:

$$\begin{pmatrix} & & & r_{1,N+1} \\ & \mathbf{R} & & \vdots \\ & & & r_{N,N+1} \\ 0 & \dots & 0 & 1 \end{pmatrix}$$

The hypothesis of "non-capture" signifies that we cannot indefinitely remain in $\mathbf{R}$ and that states $i \in [1, N]$ are called transient. That means, in particular, that for any couple $i, j \in [1, N]^2$: $\sum_{n \ge 0} p_{ij}(n) < +\infty$. Now, $p_{ij}(n) = (\mathbf{R}^n)_{ij}$; therefore, this condition indeed expresses the fact that $\mathbf{K}$ is finite.

Let us suppose we have $\boldsymbol{\pi}$ and $\tilde{q}_{ij}$ such that $\forall i, j$: $\boldsymbol{\pi}(i)\tilde{q}_{ij} = \boldsymbol{\pi}(j) q_{ji}$ and $\sum_{j \neq i} \tilde{q}_{ij} = -q_{ii}$. By definition, $\boldsymbol{\pi}$ is a stationary distribution if and only if $\boldsymbol{\pi}\mathbf{Q} = \mathbf{0}$. Let us therefore calculate $(\boldsymbol{\pi}\mathbf{Q})_i$:

$$\begin{aligned}(\boldsymbol{\pi}\mathbf{Q})_i &= \sum_j \boldsymbol{\pi}(j) q_{ji} = \boldsymbol{\pi}(i) q_{ii} + \sum_{j \neq i} \boldsymbol{\pi}(j) q_{ji} = \boldsymbol{\pi}(i) q_{ii} + \boldsymbol{\pi}(i) \sum_{j \neq i} \tilde{q}_{ij} \\ &= \boldsymbol{\pi}(i) q_{ii} - \boldsymbol{\pi}(i) q_{ii} = 0\end{aligned}$$

Thus, $\boldsymbol{\pi}$ is indeed the stationary distribution of the chain.

In the case of an M/M/c queue, the stationary distribution is given by:

$$\boldsymbol{\pi}(i) = \frac{\lambda^i}{\prod_{\mathbf{n}=1}^{i} \mu(\mathbf{n})} \boldsymbol{\pi}(0)$$

Since the queues are independent, the joint distribution $\boldsymbol{\pi}(\mathbf{n})$ is the product of the marginal distributions $\boldsymbol{\pi}_p(n_p)$; hence:

$$\boldsymbol{\pi}(\mathbf{n}) = \prod_{p=1}^{N} \boldsymbol{\pi}_p(n_p) = \prod_{p=1}^{N} \frac{\lambda^{n_p}}{\prod_{r=1}^{n_p} \mu_p(r)} \boldsymbol{\pi}_p(0)$$

According to the previously demonstrated theorem, it is enough to find $\tilde{q}_{\mathbf{n},\mathbf{m}}$ such that:

$$\forall(\mathbf{n},\mathbf{m}): \boldsymbol{\pi}(\mathbf{n})\tilde{q}_{\mathbf{n},\mathbf{m}} = \boldsymbol{\pi}(\mathbf{m})q_{\mathbf{m},\mathbf{n}} \text{ and } \textstyle\sum_{\mathbf{n}\neq\mathbf{m}} \tilde{q}_{\mathbf{n},\mathbf{m}} = -q_{\mathbf{n},\mathbf{n}} = \sum_{\mathbf{m}\neq\mathbf{n}} q_{\mathbf{n},\mathbf{m}}$$

Since all $\boldsymbol{\pi}(\mathbf{n})$ are non-zero, we can take $\tilde{q}_{ij} = \frac{\boldsymbol{\pi}(i)}{\boldsymbol{\pi}(j)} q_{ji}$ and all that is left is to verify that the two sums are equal. For that, we can expand them using the non-zero values of $q_{\mathbf{n},\mathbf{m}}$. Recall that there are three types of variations for each of the N queues in the system: exogenous arrivals, exits from the system and routing between queues. We begin by calculating the second sum because it is simpler.

$$\begin{aligned}\sum_{\mathbf{m}\neq\mathbf{n}} q_{\mathbf{n},\mathbf{m}} &= \sum_{p=1}^{N}\left(\bar{\lambda} + \mu_p(n_p)r_{p,N+1} + \sum_{q\neq p} \mu_p(n_p)r_{p,q}\right)\\ &= \sum_{p=1}^{N}\left(\bar{\lambda} + \mu_p(n_p)\sum_{q=1}^{N+1} r_{p,q}\right)\\ &= \sum_{p=1}^{N}\left(\bar{\lambda} + \mu_p(n_p)\right)\end{aligned}$$

The first sum is more technical to calculate since the roles of $\mathbf{n}$ and $\mathbf{m}$ are inverted (even if it is still $\mathbf{m}$ that varies and $\mathbf{n}$ that is constant), and we must be careful to not get lost in the indices and their meaning.

$$\begin{aligned}\sum_{\mathbf{m}\neq\mathbf{n}} \tilde{q}_{\mathbf{n},\mathbf{m}} &= \sum_{\mathbf{m}\neq\mathbf{n}} \frac{\boldsymbol{\pi}(\mathbf{m})}{\boldsymbol{\pi}(\mathbf{n})} q_{\mathbf{m},\mathbf{n}}\\ &= \sum_{p=1}^{N}\Bigg(\left[\frac{\boldsymbol{\pi}(\mathbf{n}-\mathbf{e}_p)}{\boldsymbol{\pi}(\mathbf{n})}\bar{\lambda}_p\right]_{exogenous\ arrivals}\\ &\quad + \left[\frac{\boldsymbol{\pi}(\mathbf{n}+\mathbf{e}_p)}{\boldsymbol{\pi}(\mathbf{n})}\mu_p(n_p+1)r_{p,N+1}\right]_{exits\ from\ the\ system}\\ &\quad + \left[\sum_{q\neq p} \frac{\boldsymbol{\pi}(\mathbf{n}+\mathbf{e}_q-\mathbf{e}_p)}{\boldsymbol{\pi}(\mathbf{n})}\mu_p(n_p+1)r_{q,p}\right]_{routing\ from\ q\ to\ p}\Bigg)\end{aligned}$$

$$= \sum_{p=1}^{N} \left(\frac{\mu_p(n_p)}{\lambda_p} \bar{\lambda}_p + \frac{\lambda_p}{\mu_p(n_p+1)} \mu_p(n_p+1) r_{p,N+1} \right.$$
$$\left. + \sum_{q \neq p} \frac{\lambda_q \mu_p(n_p)}{\lambda_p \mu_q(n_q+1)} \mu_q(n_q+1) r_{q,p} \right)$$
$$= \sum_{p=1}^{N} \left(\frac{\bar{\lambda}_p}{\lambda_p} \mu_p(n_p) + \lambda_p r_{p,N+1} + \frac{\mu_p(n_p)}{\lambda_p} \sum_{q \neq p} \lambda_q \, r_{q,p} \right)$$
$$= \sum_{p=1}^{N} \left(\frac{\bar{\lambda}_p}{\lambda_p} \mu_p(n_p) + \lambda_p r_{p,N+1} + \frac{\mu_p(n_p)}{\lambda_p} (\lambda_p - \bar{\lambda}_p) \right)$$
$$= \sum_{p=1}^{N} \left(\mu_p(n_p) + \lambda_p r_{p,N+1} \right)$$

The equation will therefore be satisfied if:

$$\sum_{p=1}^{N} \bar{\lambda}_p = \sum_{p=1}^{N} \lambda_p r_{p,N+1}$$

which we obtain by adding the N traffic equations since $\sum_{p \neq q} r_{q,p} = 1 - r_{q,N+1}$.

In the end, we demonstrated that even if the two hypotheses concerning the independence of queues among themselves and stating that the total arrivals in queue p follow a Poisson process with intensity λ_p have not been verified, the results are the same. This is only true for Jackson networks, and as soon as we stray from them, we quickly see we cannot obtain exact results or justify the hypotheses that we are obligated to make.

13

Chapter 5 Exercises

EXERCISE 5.1.– *On the telephone*

The total traffic corresponding to the 10 users making 10-minute phone calls is as follows:

– for an observation time of 10 minutes: $Y = 10 \times \frac{10}{10} = 10$ Erl;

– for an observation time of 30 minutes: $Y = 10 \times \frac{10}{30} = 3.33$ Erl;

– for an observation time of 1 hour: $Y = 10 \times \frac{10}{60} = 1.67$ Erl.

The traffic per user for an observation time of one hour is equal to $\frac{1.67}{10} = 0.167$ Erl, or 167 mErl.

Now, for 100 users making one-minute phone calls:

– for an observation time of 10 minutes: $Y = 100 \times \frac{1}{10} = 10$ Erl;

– for an observation time of 30 minutes: $Y = 100 \times \frac{1}{30} = 3.33$ Erl;

– for an observation time of 1 hour: $Y = 100 \times \frac{1}{60} = 1.67$ Erl.

The traffic per user for an observation time of one hour is equal to $\frac{1.67}{100} = 0.0167$ Erl, or 16.7 mErl.

These two cases have the same traffic values, whatever the observation time may be. The difference is found in the traffic per user. In the first case, the traffic per user is greater since the call duration per user is greater with respect to that of the second case.

EXERCISE 5.2.– *Road traffic*

The traffic generated by a car driving at 80 km/h is obtained from only one request, and one use of the road in $\frac{60}{80} = 0.75$ hours. We thus have: $Y = 1 \times \frac{0.75}{1} = 0.75$ Erl.

The traffic generated by 1,000 cars driving at an average speed of 80 km/h is therefore $Y = 1000 \times 0.75 = 750$ Erl.

EXERCISE 5.3.– *Business telephone lines*

The total traffic is equal to the sum of departure and arrival traffic: $0.035 + 0.045 = 0.08$ Erl. The internal traffic of 0.06 Erl is part of the total traffic of 0.08 Erl. The difference of 0.02 Erl is a sum of traffic coming from the exterior or going to the exterior.

For 10 nodes, the total traffic is equal to $10 \times 0.08 = 0.8$ Erl.

The average duration of communication is equal to: $h = Y/c = 0.08$ hour, assuming one request per hour on average. We thus have $h = 0.08 \times 60 = 4.8$ minutes, or 4 minutes 48 seconds.

EXERCISE 5.4.– *Telephone booth*

The traffic of the telephone booth: $Y = ch = (10 \times 2) \times \frac{2}{60} = 0.67$ Erl.

If the traffic is doubled, it will be $0.67 \times 2 = 1.33 > 1$, and the telephone booth can no longer carry it.

The offered traffic is equal to $A = 1.33$ Erl. The booth cannot carry more than $Y = 1$ Erl, so we have a lost (or rejected) traffic of $R = A - Y = 0.33$ Erl. This lost traffic corresponds to $c_R = R/h = \frac{0.33}{\frac{2}{60}} = 10$ refused requests on average.

On average, 10 arrivals are thus non-receivable by the booth.

EXERCISE 5.5.– *GSM*

SDCCH channels are used for four seconds when the telephone is turned on (or used by $25\% \times 100{,}000 = 25{,}000$ travelers), and for five seconds when

25% × 25,000 = 6,250 travelers make a phone call. The traffic on channel SDCCH is thus: $Y_{SDCCH} = 25000 \times \frac{4}{3600} + 6250 \times \frac{5}{3600} = 36.46$ Erl.

The TCH channels are used for 55 seconds when 6,250 travelers make a phone call. The traffic of the TCH channel is thus: $Y_{TCH} = 6250 \times \frac{55}{3600} = 95.49$ Erl.

14

Chapter 6 Exercises

EXERCISE 6.1.– *Infinite source traffic*

We have an offered traffic $A = 5$ Erl, offered by sources of infinite size. The number N of busy sources follows a Poisson distribution with parameter A.

$$\mathbb{P}(N = k) = \frac{A^k}{k!} e^{-A}$$

The probability that this traffic will be generated by only one source is: $\mathbb{P}(N = 1) = \frac{5^1}{1!} e^{-5} = 0.033$.

The probability that this traffic will be generated by five sources is: $\mathbb{P}(N = 5) = \frac{5^5}{5!} e^{-5} = 0.18$.

Five sources generate this traffic of 5 Erl for an average of one hour of activity. The arrival rate of requests is therefore five requests per hour, or $\lambda_s = 1$ request per hour per source (equal subdivision of the Poisson process). The traffic per source is equal to $A_s = \frac{5}{5} = 1$ Erl, so the average duration of activity per source is equal to $\frac{1}{\mu_s} = \frac{A_s}{\lambda_s} = 1$ hour.

The duration of activity therefore follows an exponential distribution with parameter $\mu_s = 1$.

The probability that the activity requested by one source will be greater than or equal to 30 minutes (30 minutes = 0.5 hours) is equal to $e^{-1 \times 0.5} = 0.61$.

EXERCISE 6.2.– *Busy sources*

A request arrives from one source every 10 minutes on average. With an infinite number of sources, we can therefore have six requests in one hour.

These requests arrive according to a Poisson process, so the intensity is $\lambda = 6$ requests per hour. The counting measure of the process during an interval of time Δt follows a Poisson distribution with parameter $\lambda\Delta t$.

The probability that $k = 6$ requests will arrive in $\Delta t = 1$ hour is therefore equal to: $\frac{(\lambda\Delta t)^k}{k!}e^{-\lambda\Delta t} = \frac{6^6}{6!}e^{-6} = 0.16$.

The duration of activity follows an exponential distribution with parameter μ, or with an average of $\frac{1}{\mu}$. The probability that a source will request activity for more than t is $p = e^{-\mu t}$. We deduce the value of $h = \frac{1}{\mu} = -\frac{t}{\ln p} = -\frac{30}{\ln 0.5} =$ 43.28 minutes.

The traffic offered by these sources is equal to $A = \frac{\lambda}{\mu} = c_A h = 6 \times \frac{43{,}28}{60} =$ 4.33 Erl.

The average number of busy sources is equal to 4.33 sources.

EXERCISE 6.3.– *Printer*

This time, we have a finite number of sources: the $N = 5$ employees of the small company.

Considering an exponential distribution of the durations of its use, we have $p = e^{-\mu t} = \frac{1}{3}$ for $t = 1$ minute. We deduce the average duration $h = \frac{1}{\mu} = -\frac{t}{\ln p} =$ 0.91 minutes, or 55 seconds.

For an offered traffic $A = 3$ Erl, the number of busy sources follows a binomial distribution with parameter $N = 5$ and $\frac{A}{N} = \frac{3}{5} = 0.6$. The probability that this traffic will simultaneously come from all the employees is equal to the probability that all the employees will be busy; therefore, $\binom{5}{5} \times 0.6^5 \times 0.4^0 = 0.6^5 = 0.078$.

The average number of busy employees offering this traffic of 3 Erl is equal to three employees.

The offered traffic per free employee is equal to $ch = \frac{A}{N-A} = \frac{3}{5-3} = 1.5$ Erl.

15

Chapter 7 Exercises

EXERCISE 7.1.– *PABX*

The total traffic per switchboard is equal to $0.035 + 0.045 = 0.08$ Erl. The total traffic for all of the 200 switchboards is thus $0.08 \times 200 = 16$ Erl.

The switch only has 30 circuits, so the loss probability is estimated to be 0.1% according to the Erlang-B table.

EXERCISE 7.2.– *Erlang-B table*

Table 15.1 summarizes the results for this exercise.

Loss probability	**1%**			**10%**			**50%**		
Circuits	Offered	Lost	Carried	Offered	Lost	Carried	Offered	Lost	Carried
5	1.25	0.012	1.238	3	0.3	2.7	9.5	4.75	4.75
10	4.5	0.045	4.455	7.5	0.75	6.75	18.5	9.25	9.25
15	8	0.08	7.92	12.5	1.25	11.25	28.5	14.25	14.25

Table 15.1. *Results for Exercise 7.2*

EXERCISE 7.3.– *GSM cell*

The offered traffic is equal to $A = 480 \times \frac{105}{3600} = 14$ Erl. Offering these 14 Erl to 18 circuits, the blocking rate, according to the Erlang-B table, equals 5%. For type-one pure chance traffic, the blocking rate and the congestion rate are equal; therefore, we also have a congestion rate of 5%.

Thus, 5% of the offered traffic will be lost, or $A_p = 5\% \times 14 = 0.7$ Erl. The carried traffic is therefore $Y = 13.3$ Erl, and the carried traffic per circuit is equal to $\frac{13.3}{18} = 0.74$ Erl.

The lost traffic of 0.7 Erl corresponds to a number of rejected calls per hour equal to $c_R = A_p/h = \frac{0.7}{105/3600} = 24$ rejected calls/hour.

EXERCISE 7.4.– *Authorized traffic and yield*

A circuit can carry 1 Erl. The return η of a circuit is the relationship between the traffic really carried by the circuit and the maximal traffic of 1 Erl. The Erlang-B table provided in the appendices allows us to produce Table 15.2:

Number of circuits per beam	Offered traffic	Carried traffic	η per circuit
10	6 Erl	5.7 Erl	0.57
2 x 10	12 Erl	11.4 Erl	0.57
20	15 Erl	14.25 Erl	0.71

Table 15.2. *Results for Exercise 7.4*

The smaller the line, the greater the risks of call collisions. For the same failure probability, a smaller return per circuit is observed in a small line.

EXERCISE 7.5.– *Line for outgoing calls*

The total switch capacity: the switch must carry all kinds of traffic for all phones connected to it, or $0.12 \times 120 = 14.4$ Erl.

The line for outgoing calls: the outgoing traffic to be carried first equals $0.04 \times 100 = 4$ Erl. In the case where all the circuits of the line are busy, outgoing calls are refused, and we should use the Erlang-B table. With a refusal rate of less than 10%, we find that seven circuits are needed.

EXERCISE 7.6.– *Total accessibility group (1)*

We have a finite number of sources: $N = 1\ 0$.

The offered traffic per source is equal to 0.5 Erl. Therefore, for the 10 sources, we have an offered traffic $A = 0.5 \times 10 = 5$ Erl.

The number of resources, equal to 10, is equal to the number sources, so there will be no lost traffic.

The carried traffic is equal to the offered traffic: $Y = 5$ Erl. The lost traffic $A_p = 0$, and the blocking rate is $B = 0$.

The congestion rate can be calculated using formula [7.23] where the offered traffic per free source is equal to $b = \frac{A}{N-A} = \frac{5}{10-5} = 1$ Erl:

$$E = E_{10.10}(1) = \frac{\binom{10}{10} 1^{10}}{\sum_{i=0}^{10} \binom{10}{i} 1^i} = 0.098\%$$

EXERCISE 7.7.– *Total accessibility group (2)*

We have a large number of sources, which we will consider infinite.

The offered traffic is equal to $A = 5$ Erl. With 10 resources, an offered traffic of 5 Erl gives us a blocking rate of 5% (see the Erlang-B table in Appendix 2). For Poissonian traffic, the congestion rate is equal to the blocking rate, or 5%.

The lost traffic is therefore equal to 5% × 5 = 0.25 Erl. The carried traffic equals 5 – 0.25 = 4.75 Erl.

The Erlang model uses a greater number of resources offering the traffic of 5 Erl; the congestion rate is higher than that of the Engset model in Exercise 7.6.

EXERCISE 7.8.– *Call center*

The average duration of a call is 30 seconds, or a processing capacity of two calls/minute. A line (1 Erl) is therefore capable of receiving 120 calls/hour. The traffic to be carried is 360,000 calls/hour (720,000/2).

The number of lines needed is: 360,000/120 = 3,000 lines, or 3,000 Erl.

EXERCISE 7.9.– *Back to the past of the X.25*

We must first determine the traffic that will be carried and then define the number of circuits needed using the Erlang-B table.

– traffic to be carried: A= 600 × 2 /60 = 20 Erl;

– number of virtual circuits for a quality of service that is better than 1%: the table provides the responses of 30 circuits;

– number of satisfied requests for a refusal rate of 2%: with 30 circuits and a refusal rate of 2%, the offered traffic is equal to 22 Erl. The lost traffic is equal to 2% × 22 = 0.44 Erl. The carried traffic is equal to 22 – 0.44 = 21.56 Erl. The number of satisfied requests can be obtained from $c_Y = \frac{Y}{h} = \frac{21.56}{2/60} = 646$ requests.

16

Chapter 8 Exercises

EXERCISE 8.1.– *Total accessibility with delay (1)*

The offered traffic is equal to $A = 700 \times \frac{108}{3600} = 21$ Erl.

Looking at the Erlang-C table, the probability of waiting corresponding to 30 circuits and an offered traffic of 21 Erl is equal to $D = 5\%$.

The carried traffic is equal to the offered traffic since there is no loss, so $Y =$ 21 Erl. Thus, the average carried traffic per channel is equal to $\frac{21}{30} = 0.70$ Erl.

The average delay time for all the calls is given by:

$$W = E(d) = D.\frac{h}{n-A} = 0.05.\frac{108}{30-21} = 0.6 \text{ seconds.}$$

The average delay time for calls that must wait is equal to:

$$w = E(d_q) = \frac{h}{n-A} = \frac{108}{30-21} = 12 \text{ seconds.}$$

EXERCISE 8.2.– *Total accessibility with delay (2)*

Calls are served in the order of their arrival, so the delay time follows an exponential distribution . The probability that a call will have to wait for more than t seconds is given by: $\mathbb{P}(d > t) = D.e^{-\frac{t}{E(d_q)}} = D.e^{-\frac{t(n-A)}{h}} = 0.05.e^{-\frac{t}{12}}$. Calculations give 0.04, 0.03 and 0.02, respectively, as the probabilities that a call will have to wait more than 3, 6 and 12 seconds.

It is the same for the probability of delay times given that a call must wait: $\mathbb{P}(d_q > t) = e^{-\frac{t}{E(d_q)}} = e^{-\frac{t(n-A)}{h}} = e^{-\frac{t}{12}}$. Calculations give 0.78, 0.61 and 0.37, respectively, as the probabilities that a call will have to wait more than 3, 6 and 12 seconds, given that it has to wait.

EXERCISE 8.3.– *Phone-in*

The probability that a call must wait more than 3 seconds is expressed by $p = \mathbb{P}(d > t) = De^{-\frac{t}{E(d_q)}} = De^{-\frac{t(n-A)}{h}}$, where $A = 30 \times \frac{h}{3600}$ is the offered traffic, $n = 1$ resource and D is the probability of waiting.

First note that $D = A$ for the case $n = 1$. Indeed, we have:

$$D = \frac{\frac{A^n}{n!}\frac{n}{n-A}}{1 + A + \frac{A^2}{2!} + \cdots + \frac{A^{n-1}}{(n-1)!} + \frac{A^n}{n!}\cdot\frac{n}{n-A}} = \frac{\frac{A}{1-A}}{1 + \frac{A}{1-A}} = A$$

It follows that:

$$\mathbb{P}(d > t) = Ae^{-\frac{t(n-A)}{h}} = \frac{h}{120}e^{-\frac{t\left(1-\frac{h}{120}\right)}{h}}$$

This probability should not be greater than 0.1 for $t = 3$ seconds.

Thus: $0.1 > \frac{h}{120}e^{-\frac{3(1-h/120)}{h}} = \frac{h}{120}e^{-\frac{3}{h}+\frac{3}{120}}$, $12e^{-\frac{3}{120}} > he^{-\frac{3}{h}}$, $h < 14.5$. (We can use the approximation $e^{-\frac{3}{h}} \simeq 1 - \frac{3}{h}$ for a small $\frac{3}{h}$, and we reach the approximate result of $h < 14.7$.)

EXERCISE 8.4.– *Delay*

The probability p that an incoming call will encounter another call in the queue that has been waiting for more than 30 seconds is equal to the probability that the delay time will be greater than 30 seconds given that the call had to wait. This probability is equal to $p = \mathbb{P}(d_q > t) = e^{-\frac{t}{E(d_q)}} = e^{-\frac{t(n-A)}{h}}$ for $n = 10$, $A = 8$ Erl and $h = 5$ minutes $= 5 \times 60$ seconds. We therefore have:

$$p = e^{-\frac{30(10-8)}{5\times 60}} = 0.82 \text{ seconds}$$

EXERCISE 8.5.– *Line for incoming calls*

First refer to Exercise 7.5.

Incoming call line: the delay time is less than 2%. If the interlocutor is unavailable, no traffic will be carried while the caller occupies a circuit (waiting music). In these conditions, we must use the waiting table (Erlang-C).

Offered traffic: $A = 0.004 \times 100 = 4$ Erl.

The Erlang-C table for delay times in Appendix 2 indicates that nine circuits are needed.

EXERCISE 8.6.– *Constant holding*

The probability that an incoming request will have to wait is equal to the probability that the inter-arrival of requests will be less than the duration of processing $h = 1$ microsecond. In this case, the first request has not yet finished with its service, but the other has already arrived, so it must wait.

This probability is equal to 0.23. Indeed, arrivals form a Poisson process with a rate of $\lambda = 4{,}320$ requests per minute $= 4{,}320 \times 60$ requests per second. The inter-arrival τ follows an exponential distribution with parameter λ, so:

$$\mathbb{P}(\tau < 3 \times 10^{-6}) = 1 - e^{-4320 \times 60 \times 3 \times 10^{-6}} = 0.54$$

The offered traffic is equal to $A = 4320 \times 60 \times 3 \times 10^{-6} = 0.78$. The average delay time for requests that must wait is equal to: $w = \frac{h}{2(1-A)} = \frac{3}{2(1-0.78)} =$ 6.82 microseconds.

The average delay time for all of the requests is equal to: $W = \frac{hA}{2(1-A)} = \frac{3 \times 0.78}{2(1-0.78)} = 5.32$ microseconds.

The average response time of the server is equal to 8.32 microseconds. We use $E(T_s) = E(T_q) + h$, where $E(T_q) = W$.

EXERCISE 8.7.– *Terminals in a distribution chain*

The number of checkout terminals: Starting with the delay time ($W = E(d)$) to reach the registers (10 min), we determine the maximal possible load ($\rho = A$). From this load, given the time it takes to check out (duration of service or h), we define the number of customers processed per checkout terminal while keeping in mind that for $n = 1, D = A$.

$$W = E(d) = A\frac{h}{1-A}, \qquad A = \frac{W}{W+h} = \frac{10}{10+3} = 0.76$$

A maximal load of 0.76 corresponds to an arrival rate to the checkout terminal of:

$$A = c_A h, \quad c_A = \frac{A}{h} = 0.76 \times \frac{60}{3} = 15\ customers/hour$$

The number of customers in the queue is:

$$E(N_q) = c_A E(d) = 15 \times 10/60 = 2.5\ customers$$

What corresponds to the specifications, in these conditions, is to have four checkout terminals to carry the traffic of 60 customers/hour.

The number of collection point terminals: The collection terminals must be accessible in 80% of cases, that is, a salesclerk must not refuse access to the terminal in more than 20% of cases. The number of terminals can be defined using the Erlang-B table. Given that the service occupies the terminal for one minute and that 60 customers come within the space of an hour, the traffic to be carried is equal to: $A = c_A\frac{h}{60} = 60.\frac{1}{60} = 1$ Erl. The table indicates two terminals.

The number of sales point terminals: The reasoning is identical. Given that there are 100 consultations of one minute and 60 orders taking 3 minutes, the traffic to be carried is: $A = c_A\frac{h}{60} = \frac{100 \cdot 1 + 60 \cdot 3}{60} = 4.66$ Erl. The table, for a refusal rate of 5%, gives eight terminals.

The number of terminals that need to be installed is presented in Table 16.1:

Terminals	Sales point	Register	Collection point	Accounting
Number	8	4	2	2

Table 16.1. *Number of terminals to install – Exercise 8.7*

Response time: The response time expresses the operator delay time from the moment when they validate a request to when the response is displayed. Figure 16.1 sets out the components of this duration.

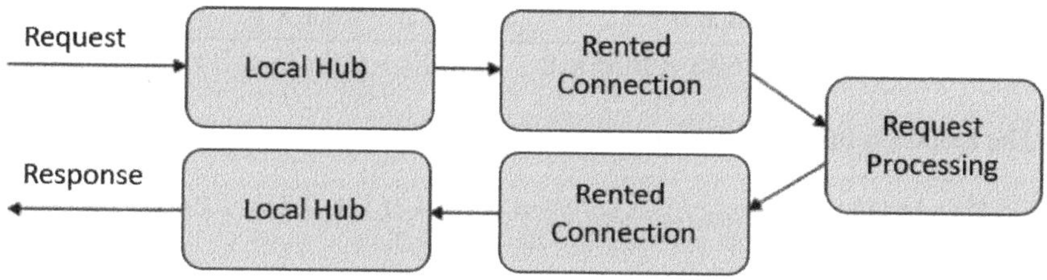

Figure 16.1. *Definition of response time – Exercise 8.7*

The data transport time is determined from the average transaction. Recall that there are 100 transactions/hour at the sales point, 60 transactions/hour at the register and collections and 40 transactions/day (or 5 transactions/hour) for the accounting terminals. The average transaction must be defined in the sense of the *host/branch* (LHS) and in the sense of the *branch/host* (LSH), and the average length must be multiplied by 1.2 to account for service data:

$$L_{HS} = \frac{\sum \lambda L}{\sum L} = \frac{100 \times 800 + 60 \times 600 + 60 \times 500 + 5 \times 800}{100 + 60 + 60 + 5}.1.2 = 800\ bytes$$

$$L_{SH} = \frac{\sum \lambda L}{\sum L} = \frac{100 \times 20 + 60 \times 100 + 60 \times 20 + 5 \times 200}{100 + 60 + 60 + 5}.1.2 = 54.4\ bytes$$

The number of transactions or arrival rate is:

$$\lambda = \sum \lambda_i = 100 + 60 + 60 + 5 = 225\, trans/hour = 0.0625\ trans/sec$$

Response time of the local hub (t_{rc}), considering the time it takes to transfer data from the hub to the terminals (request and response):

$$t_{rc} = \frac{L_{HS}}{D_c - \lambda L_{HS}} + \frac{L_{SH}}{D_c - \lambda L_{SH}}$$
$$= \frac{800 \times 8}{9600 - 0.0625 \times 800 \times 8} + \frac{54.4 \times 8}{9600 - 0.0625 \times 54.4 \times 8} = 0.74\ s$$

The response time of the rented connection (t_{rl}) is:

$$t_{rl} = \frac{L_{HS}}{D_t - \lambda L_{HS}} + \frac{L_{SH}}{D_t - \lambda L_{SH}}$$
$$= \frac{800 \times 8}{64000 - 0.0625 \times 800 \times 8} + \frac{54.4 \times 8}{64000 - 0.0625 \times 54.4 \times 8} = 0.10\ s$$

The response time of the transaction is:

$$Tr = t_{rc} + t_{rl} + t_{processing_{request}} = 0.74 + 0.1 + 0.2 = 1.04\ s$$

Recall that when the system has a light load, which is often the case in conventional systems, we can simply say: $Tr = L/D$.

EXERCISE 8.8.– *Teletraffic in a local area network*

Arrival rate of packets λ: the arrival rate of packets is determined by applying the principle of superposed flows, or:

$$\lambda = 2 \times 4 + 2 \times 2 + 3 \times 6 + 5 \times 5 = 55\ packets/s$$

Service rate μ: The service rate represents the number of packets processed per second. It is given by the equation $\mu = 1/h$, where h represents the duration of service, or:

$$h = \left(128\ \times 8 \times \frac{1}{64000}\right) = 16\ ms$$

$$\mu = 1/16\,.\,10^{-3} = 62.5\ packets/second$$

System load ρ: The system load or traffic intensity is the relationship between the submitted load and the admissible load: $\rho = 55/62.5 = 0.88$ Erl. Note that the system is stable since $\rho < 1$, but close to saturation.

Number of packets in the router N: the number of packets in the router is given by the equation:

$$N = \frac{\rho}{1-\rho} = \frac{0.88}{1 - 0.88} = 7.3\ packets$$

Average delay time W: The average delay time corresponds to the product of the number of packets in the router and the processing time of one packet (duration of service), or: $W = \frac{\rho h}{1-\rho} = 7.3 \times \left(\frac{128\times 8}{64000}\right) = 7.3 \times 0.016 = 0.1168$ seconds.

Number of packets in the queue (waiting packets N_f): $N_f = \lambda \times W = 55 \times 0.1168 = 6.424\ packets$

Response time or queuing time w: $w = \frac{h}{1-\rho} = \frac{0.016}{1-0.88} = 0.1333\ s$

Buffer time T: $T = Number\ of\ waiting\ packets\ \times size\ of\ the\ packet = 6.424\ \times 128 = 822\ bytes$

If we use a buffer size of 1 kilobyte or a capacity of eight packets, the queue is therefore M/M/1/8.

Loss probability of one packet: the average number of packets in the system is 7.3 packets.

If $\neq 1$:

$$p_n = \frac{\rho^n(1-\rho)}{1-\rho^{K+1}} = \frac{0.88^{7.3}(1-0.88)}{1-0.88^{8+1}} = 0.075$$

Note that for a load of 50%, the loss probability would still be 2% and therein lies the difficulty of dimensioning temp memory in active elements.

EXERCISE 8.9.– *Teletraffic in a computer network*

The method consists of considering the network one queue and Little's law to determine the transit time. For that, we should, using the carried traffic for each node, determine the number of packets in transit in the network and then apply Little's law.

Let us recall the particulars:

– average length of a packet: 128 bytes;

– bandwidth of connections: 64 kbit/s;

– arrival rate in E: $\lambda = 30$ packets/s;

– service time: $h = 128 \times 8/64{,}000 = 16$ ms;

– service rate: $\mu = 1/h = 1/(16 \times 10^{-3}) = 62.5$ packets/s.

Table 16.2 summarizes the figures.

File	Traffic %	Arrival rate λ	Load $\rho = \lambda/\mu$	Number of items in node $N = \rho/(1-\rho)$
F1:2	0.75	22.5	0.36	0.5625
F1:3	0.25	7.5	0.12	0.1363
F2:4	0.375	11.25	0.18	0.2195
F2:5	0.375	11.25	0.18	0.2195
F3:4	0.1875	5.625	0.09	0.098
F3:5	0.0625	1.875	0.03	0.0309
F4:6	0.5625	16.875	0.27	0.3698
F5:6	0.4375	13.125	0.21	0.2658
F6:S	1	30	0.48	0.9230
Number of packets in the network				2.8253

Table 16.2. *Summary of Exercise 8.9*

The transit time in the network is equal to $W + h$. $N = \lambda \times W$ or $W = N/\lambda = 2{,}8253/30 = 94$ ms, and $h = 16 \times 4 = 64$ ms, so the transit time is equal to 158 ms.

PART 5

Appendices

Appendix 1

Erlang-B Table

Offered traffic A relative to the number of resources N and the loss probability B (%).

N/B	0.01%	0.05%	0.1%	0.5%	1.0%	2%	5%	10%	15%	20%	30%	40%
1	0.000	0.000	0.001	0.005	0.010	0.020	0.052	0.111	0.176	0.250	0.428	0.666
2	0.014	0.032	0.045	0.105	0.152	0.223	0.381	0.595	0.796	1.000	1.449	2.000
3	0.086	0.151	0.193	0.349	0.455	0.602	0.899	1.271	1.603	1.930	2.633	3.480
4	0.234	0.362	0.439	0.701	0.869	1.092	1.525	2.045	2.501	2.945	3.891	5.021
5	0.452	0.648	0.762	1.132	1.361	1.657	2.219	2.881	3.454	4.010	5.189	6.596
6	0.728	0.995	1.146	1.622	1.909	2.276	2.960	3.758	4.445	5.109	6.514	8.191
7	1.054	1.392	1.579	2.158	2.501	2.935	3.738	4.666	5.461	6.230	7.856	9.800
8	1.422	1.830	2.051	2.730	3.128	3.627	4.543	5.597	6.498	7.369	9.213	11.42
9	1.826	2.302	2.558	3.333	3.783	4.345	5.370	6.546	7.551	8.522	10.58	13.05
10	2.260	2.803	3.092	3.961	4.461	5.084	6.216	7.511	8.616	9.685	11.95	14.68
11	2.722	3.329	3.651	4.610	5.160	5.842	7.076	8.487	9.691	10.86	13.33	16.31
12	3.207	3.878	4.231	5.279	5.876	6.615	7.950	9.474	10.78	12.04	14.72	17.95
13	3.713	4.447	4.831	5.964	6.607	7.402	8.835	10.47	11.87	13.22	16.11	19.60

N/B	0.01%	0.05%	0.1%	0.5%	1.0%	2%	5%	10%	15%	20%	30%	40%
14	4.239	5.032	5.446	6.663	7.352	8.200	9.730	11.47	12.97	14.41	17.50	21.24
15	4.781	5.634	6.077	7.376	8.108	9.010	10.63	12.48	14.07	15.61	18.90	22.89
16	5.339	6.250	6.722	8.100	8.875	9.828	11.54	13.50	15.18	16.81	20.30	24.54
17	5.911	6.878	7.378	8.834	9.652	10.66	12.46	14.52	16.29	18.01	21.70	26.19
18	6.496	7.519	8.046	9.578	10.44	11.49	13.39	15.55	17.41	19.22	23.10	27.84
19	7.093	8.170	8.724	10.33	11.23	12.33	14.32	16.58	18.53	20.42	24.51	29.50
20	7.701	8.831	9.412	11.09	12.03	13.18	15.25	17.61	19.65	21.64	25.92	31.15
21	8.319	9.501	10.11	11.86	12.84	14.04	16.19	18.65	20.77	22.85	27.33	32.81
22	8.946	10.18	10.81	12.64	13.65	14.90	17.13	19.69	21.90	24.06	28.74	34.46
23	9.583	10.87	11.52	13.42	14.47	15.76	18.08	20.74	23.03	25.28	30.15	36.12
24	10.23	11.56	12.24	14.20	15.30	16.63	19.03	21.78	24.16	26.50	31.56	37.78
25	10.88	12.26	12.97	15.00	16.13	17.51	19.99	22.83	25.30	27.72	32.97	39.44
26	11.54	12.97	13.70	15.80	16.96	18.38	20.94	23.89	26.43	28.94	34.39	41.10
27	12.21	13.69	14.44	16.60	17.80	19.27	21.90	24.94	27.57	30.16	35.80	42.76
28	12.88	14.41	15.18	17.41	18.64	20.15	22.87	26.00	28.71	31.39	37.21	44.41
29	13.56	15.13	15.93	18.22	19.49	21.04	23.83	27.05	29.85	32.61	38.63	46.07
30	14.25	15.86	16.68	19.03	20.34	21.93	24.80	28.11	31.00	33.84	40.05	47.74
31	14.94	16.60	17.44	19.85	21.19	22.83	25.77	29.17	32.14	35.07	41.46	49.40
32	15.63	17.34	18.21	20.68	22.05	23.73	26.75	30.24	33.28	36.30	42.88	51.06
33	16.34	18.09	18.97	21.51	22.91	24.63	27.72	31.30	34.43	37.52	44.30	52.72
34	17.04	18.84	19.74	22.34	23.77	25.53	28.70	32.37	35.58	38.75	45.72	54.38
35	17.75	19.59	20.52	23.17	24.64	26.44	29.68	33.43	36.72	39.99	47.14	56.04

N/B	0.01%	0.05%	0.1%	0.5%	1.0%	2%	5%	10%	15%	20%	30%	40%
36	18.47	20.35	21.30	24.01	25.51	27.34	30.66	34.50	37.87	41.22	48.56	57.70
37	19.19	21.11	22.08	24.85	26.38	28.25	31.64	35.57	39.02	42.45	49.98	59.37
38	19.91	21.87	22.86	25.69	27.25	29.17	32.62	36.64	40.17	43.68	51.40	61.03
39	20.64	22.64	23.65	26.53	28.13	30.08	33.61	37.72	41.32	44.91	52.82	62.69
40	21.37	23.41	24.44	27.38	29.01	31.00	34.60	38.79	42.48	46.15	54.24	64.35
41	22.11	24.19	25.24	28.23	29.89	31.92	35.58	39.86	43.63	47.38	55.66	66.02
42	22.85	24.97	26.04	29.09	30.77	32.84	36.57	40.94	44.78	48.62	57.08	67.68
43	23.59	25.75	26.84	29.94	31.66	33.76	37.57	42.01	45.94	49.85	58.50	69.34
44	24.33	26.53	27.64	30.80	32.54	34.68	38.56	43.09	47.09	51.09	59.92	71.01
45	25.08	27.32	28.45	31.66	33.43	35.61	39.55	44.17	48.25	52.32	61.35	72.67
46	25.83	28.11	29.26	32.52	34.32	36.53	40.55	45.24	49.40	53.56	62.77	74.33
47	26.59	28.90	30.07	33.38	35.22	37.46	41.54	46.32	50.56	54.80	64.19	76.00
48	27.34	29.70	30.88	34.25	36.11	38.39	42.54	47.40	51.71	56.03	65.61	77.66
49	28.10	30.49	31.69	35.11	37.00	39.32	43.53	48.48	52.87	57.27	67.04	79.32
50	28.87	31.29	32.51	35.98	37.90	40.26	44.53	49.56	54.03	58.51	68.46	80.99
51	29.63	32.09	33.33	36.85	38.80	41.19	45.53	50.64	55.19	59.75	69.88	82.65
52	30.40	32.90	34.15	37.72	39.70	42.12	46.53	51.73	56.35	60.99	71.31	84.32
53	31.17	33.70	34.98	38.60	40.60	43.06	47.53	52.81	57.50	62.22	72.73	85.98
54	31.94	34.51	35.80	39.47	41.51	44.00	48.54	53.89	58.66	63.46	74.15	87.65
55	32.72	35.32	36.63	40.35	42.41	44.94	49.54	54.98	59.82	64.70	75.58	89.31
56	33.49	36.13	37.46	41.23	43.32	45.88	50.54	56.06	60.98	65.94	77.00	90.97
57	34.27	36.95	38.29	42.11	44.22	46.82	51.55	57.14	62.14	67.18	78.43	92.64

N/B	0.01%	0.05%	0.1%	0.5%	1.0%	2%	5%	10%	15%	20%	30%	40%
58	35.05	37.76	39.12	42.99	45.13	47.76	52.55	58.23	63.31	68.42	79.85	94.30
59	35.84	38.58	39.96	43.87	46.04	48.70	53.56	59.32	64.47	69.66	81.27	95.97
60	36.62	39.40	40.80	44.76	46.95	49.64	54.57	60.40	65.63	70.90	82.70	97.63
61	37.41	40.22	41.63	45.64	47.86	50.59	55.57	61.49	66.79	72.14	84.12	99.30
62	38.20	41.05	42.47	46.53	48.77	51.53	56.58	62.58	67.95	73.38	85.55	101.0
63	38.99	41.87	43.31	47.42	49.69	52.48	57.59	63.66	69.11	74.63	86.97	102.6
64	39.78	42.70	44.16	48.31	50.60	53.43	58.60	64.75	70.28	75.87	88.40	104.3
65	40.58	43.52	45.00	49.20	51.52	54.38	59.61	65.84	71.44	77.11	89.82	106.0
66	41.38	44.35	45.85	50.09	52.44	55.33	60.62	66.93	72.60	78.35	91.25	107.6
67	42.17	45.18	46.69	50.98	53.35	56.28	61.63	68.02	73.77	79.59	92.67	109.3
68	42.97	46.02	47.54	51.87	54.27	57.23	62.64	69.11	74.93	80.83	94.10	111.0
69	43.77	46.85	48.39	52.77	55.19	58.18	63.65	70.20	76.09	82.08	95.52	112.6
70	44.58	47.68	49.24	53.66	56.11	59.13	64.67	71.29	77.26	83.32	96.95	114.3
71	45.38	48.52	50.09	54.56	57.03	60.08	65.68	72.38	78.42	84.56	98.37	116.0
72	46.19	49.36	50.94	55.46	57.96	61.04	66.69	73.47	79.59	85.80	99.80	117.6
73	47.00	50.20	51.80	56.35	58.88	61.99	67.71	74.56	80.75	87.05	101.2	119.3
74	47.81	51.04	52.65	57.25	59.80	62.95	68.72	75.65	81.92	88.29	102.7	120.9
75	48.62	51.88	53.51	58.15	60.73	63.90	69.74	76.74	83.08	89.53	104.1	122.6
76	49.43	52.72	54.37	59.05	61.65	64.86	70.75	77.83	84.25	90.78	105.5	124.3
77	50.24	53.56	55.23	59.96	62.58	65.81	71.77	78.93	85.41	92.02	106.9	125.9
78	51.05	54.41	56.09	60.86	63.51	66.77	72.79	80.02	86.58	93.26	108.4	127.6
79	51.87	55.25	56.95	61.76	64.43	67.73	73.80	81.11	87.74	94.51	109.8	129.3

N/B	0.01%	0.05%	0.1%	0.5%	1.0%	2%	5%	10%	15%	20%	30%	40%
80	52.69	56.10	57.81	62.67	65.36	68.69	74.82	82.20	88.91	95.75	111.2	130.9
81	53.51	56.95	58.67	63.57	66.29	69.65	75.84	83.30	90.08	96.99	112.6	132.6
82	54.33	57.80	59.54	64.48	67.22	70.61	76.86	84.39	91.24	98.24	114.1	134.3
83	55.15	58.65	60.40	65.39	68.15	71.57	77.87	85.48	92.41	99.48	115.5	135.9
84	55.97	59.50	61.27	66.29	69.08	72.53	78.89	86.58	93.58	100.7	116.9	137.6
85	56.79	60.35	62.14	67.20	70.02	73.49	79.91	87.67	94.74	102.0	118.3	139.3
86	57.62	61.21	63.00	68.11	70.95	74.45	80.93	88.77	95.91	103.2	119.8	140.9
87	58.44	62.06	63.87	69.02	71.88	75.42	81.95	89.86	97.08	104.5	121.2	142.6
88	59.27	62.92	64.74	69.93	72.82	76.38	82.97	90.96	98.25	105.7	122.6	144.3
89	60.10	63.77	65.61	70.84	73.75	77.34	83.99	92.05	99.41	107.0	124.0	145.9
90	60.92	64.63	66.48	71.76	74.68	78.31	85.01	93.15	100.6	108.2	125.5	147.6
91	61.75	65.49	67.36	72.67	75.62	79.27	86.04	94.24	101.8	109.4	126.9	149.3
92	62.58	66.35	68.23	73.58	76.56	80.24	87.06	95.34	102.9	110.7	128.3	150.9
93	63.42	67.21	69.10	74.50	77.49	81.20	88.08	96.43	104.1	111.9	129.8	152.6
94	64.25	68.07	69.98	75.41	78.43	82.17	89.10	97.53	105.3	113.2	131.2	154.3
95	65.08	68.93	70.85	76.33	79.37	83.13	90.12	98.63	106.4	114.4	132.6	155.9
96	65.92	69.79	71.73	77.24	80.31	84.10	91.15	99.72	107.6	115.7	134.0	157.6
97	66.75	70.65	72.61	78.16	81.25	85.07	92.17	100.8	108.8	116.9	135.5	159.3
98	67.59	71.52	73.48	79.07	82.18	86.04	93.19	101.9	109.9	118.2	136.9	160.9
99	68.43	72.38	74.36	79.99	83.12	87.00	94.22	103.0	111.1	119.4	138.3	162.6
100	69.27	72.5	75.24	80.91	84.06	87.97	95.24	104.1	112.3	120.6	139.7	164.3

Appendix 2

Erlang-C Table

Offered traffic A relative to the number of resources N and the probability of waiting D (%).

N/D	0.01%	0.05%	0.1%	0.5%	1.0%	2%	5%	10%	15%	20%	30%	40%
1	0.000	0.000	0.001	0.005	0.010	0.020	0.050	0.100	0.150	0.200	0.300	0.400
2	0.014	0.031	0.045	0.102	0.146	0.210	0.342	0.500	0.627	0.740	0.939	1.117
3	0.086	0.149	0.189	0.333	0.429	0.554	0.787	1.040	1.231	1.393	1.667	1.903
4	0.231	0.353	0.425	0.664	0.810	0.993	1.319	1.653	1.899	2.102	2.440	2.725
5	0.442	0.628	0.734	1.065	1.259	1.497	1.905	2.313	2.607	2.847	3.241	3.569
6	0.711	0.961	1.099	1.519	1.758	2.047	2.532	3.007	3.344	3.617	4.062	4.428
7	1.026	1.341	1.510	2.014	2.297	2.633	3.188	3.725	4.103	4.406	4.897	5.298
8	1.382	1.758	1.958	2.543	2.866	3.246	3.869	4.463	4.878	5.210	5.744	6.178
9	1.771	2.208	2.436	3.100	3.460	3.883	4.569	5.218	5.668	6.027	6.600	7.065
10	2.189	2.685	2.942	3.679	4.077	4.540	5.285	5.986	6.469	6.853	7.465	7.959
11	2.634	3.186	3.470	4.279	4.712	5.213	6.015	6.765	7.280	7.688	8.336	8.857
12	3.100	3.708	4.018	4.896	5.363	5.901	6.758	7.554	8.099	8.530	9.212	9.761
13	3.587	4.248	4.584	5.529	6.028	6.602	7.511	8.352	8.926	9.379	10.09	10.67

N/D	0.01%	0.05%	0.1%	0.5%	1.0%	2%	5%	10%	15%	20%	30%	40%
14	4.092	4.805	5.166	6.175	6.705	7.313	8.273	9.158	9.760	10.23	10.98	11.58
15	4.614	5.377	5.762	6.833	7.394	8.035	9.044	9.970	10.60	11.09	11.87	12.49
16	5.150	5.962	6.371	7.502	8.093	8.766	9.822	10.79	11.44	11.96	12.77	13.41
17	5.699	6.560	6.991	8.182	8.801	9.505	10.61	11.61	12.29	12.83	13.66	14.33
18	6.261	7.169	7.622	8.871	9.518	10.25	11.40	12.44	13.15	13.70	14.56	15.25
19	6.835	7.788	8.263	9.568	10.24	11.01	12.20	13.28	14.01	14.58	15.47	16.18
20	7.419	8.417	8.914	10.27	10.97	11.77	13.00	14.12	14.87	15.45	16.37	17.10
21	8.013	9.055	9.572	10.99	11.71	12.53	13.81	14.96	15.73	16.34	17.28	18.03
22	8.616	9.702	10.24	11.70	12.46	13.30	14.62	15.81	16.60	17.22	18.19	18.96
23	9.228	10.36	10.91	12.43	13.21	14.08	15.43	16.65	17.47	18.11	19.10	19.89
24	9.848	11.02	11.59	13.16	13.96	14.86	16.25	17.51	18.35	19.00	20.02	20.82
25	10.48	11.69	12.28	13.90	14.72	15.65	17.08	18.36	19.22	19.89	20.93	21.76
26	11.11	12.36	12.97	14.64	15.49	16.44	17.91	19.22	20.10	20.79	21.85	22.69
27	11.75	13.04	13.67	15.38	16.26	17.23	18.74	20.08	20.98	21.68	22.77	23.63
28	12.40	13.73	14.38	16.14	17.03	18.03	19.57	20.95	21.87	22.58	23.69	24.57
29	13.05	14.42	15.09	16.89	17.81	18.83	20.41	21.82	22.75	23.48	24.61	25.50
30	13.71	15.12	15.80	17.65	18.59	19.64	21.25	22.68	23.64	24.38	25.54	26.44
31	14.38	15.82	16.52	18.42	19.37	20.45	22.09	23.56	24.53	25.29	26.46	27.38
32	15.05	16.53	17.25	19.18	20.16	21.26	22.93	24.43	25.42	26.19	27.39	28.33
33	15.72	17.24	17.97	19.95	20.95	22.07	23.78	25.30	26.32	27.10	28.31	29.27
34	16.40	17.95	18.71	20.73	21.75	22.89	24.63	26.18	27.21	28.01	29.24	30.21
35	17.09	18.67	19.44	21.51	22.55	23.71	25.48	27.06	28.11	28.92	30.17	31.16

N/D	0.01%	0.05%	0.1%	0.5%	1.0%	2%	5%	10%	15%	20%	30%	40%
36	17.78	19.39	20.18	22.29	23.35	24.53	26.34	27.94	29.00	29.83	31.10	32.10
37	18.47	20.12	20.92	23.07	24.15	25.36	27.19	28.82	29.90	30.74	32.03	33.05
38	19.17	20.85	21.67	23.86	24.96	26.18	28.05	29.71	30.80	31.65	32.97	34.00
39	19.87	21.59	22.42	24.65	25.77	27.01	28.91	30.59	31.71	32.57	33.90	34.94
40	20.58	22.33	23.17	25.44	26.58	27.84	29.77	31.48	32.61	33.48	34.83	35.89
41	21.28	23.07	23.93	26.23	27.39	28.68	30.63	32.37	33.51	34.40	35.77	36.84
42	22.00	23.81	24.69	27.03	28.21	29.51	31.50	33.26	34.42	35.32	36.70	37.79
43	22.71	24.56	25.45	27.83	29.02	30.35	32.36	34.15	35.33	36.23	37.64	38.74
44	23.43	25.31	26.22	28.63	29.84	31.19	33.23	35.04	36.23	37.15	38.58	39.69
45	24.15	26.06	26.98	29.44	30.67	32.03	34.10	35.93	37.14	38.07	39.51	40.64
46	24.88	26.82	27.75	30.24	31.49	32.87	34.97	36.83	38.05	39.00	40.45	41.59
47	25.60	27.57	28.52	31.05	32.32	33.72	35.84	37.72	38.96	39.92	41.39	42.54
48	26.34	28.33	29.30	31.86	33.14	34.56	36.72	38.62	39.87	40.84	42.33	43.50
49	27.07	29.10	30.08	32.68	33.97	35.41	37.59	39.52	40.79	41.76	43.27	44.45
50	27.80	29.86	30.86	33.49	34.80	36.26	38.47	40.42	41.70	4?.69	44.21	45.40
51	28.54	30.63	31.64	34.31	35.64	37.11	39.35	41.32	42.61	43.61	45.15	46.36
52	29.28	31.40	32.42	35.12	36.47	37.97	40.23	42.22	43.53	44.54	46.10	47.31
53	30.03	32.17	33.21	35.94	37.31	38.82	41.10	43.12	44.44	45.47	47.04	48.27
54	30.77	32.95	33.99	36.76	38.15	39.67	41.99	44.02	45.36	46.39	47.98	49.22
55	31.52	33.72	34.78	37.59	38.99	40.53	42.87	44.93	46.28	47.32	48.93	50.18
56	32.27	34.50	35.57	38.41	39.83	41.39	43.75	45.83	47.20	48.25	49.87	51.13
57	33.03	35.28	36.37	39.24	40.67	42.25	44.64	46.74	48.12	49.18	50.82	52.09

N/D	0.01%	0.05%	0.1%	0.5%	1.0%	2%	5%	10%	15%	20%	30%	40%
58	33.78	36.06	37.16	40.07	41.51	43.11	45.52	47.64	49.04	50.11	51.76	53.05
59	34.54	36.85	37.96	40.90	42.36	43.97	46.41	48.55	49.96	51.04	52.71	54.01
60	35.30	37.63	38.76	41.73	43.20	44.83	47.29	49.46	50.88	51.97	53.65	54.96
61	36.06	38.42	39.56	42.56	44.05	45.70	48.18	50.37	51.80	52.90	54.60	55.92
62	36.82	39.21	40.36	43.39	44.90	46.56	49.07	51.27	52.72	53.83	55.55	56.88
63	37.59	40.00	41.16	44.23	45.75	47.43	49.96	52.18	53.64	54.77	56.49	57.84
64	38.35	40.80	41.97	45.06	46.60	48.30	50.85	53.10	54.57	55.70	57.44	58.80
65	39.12	41.59	42.78	45.90	47.45	49.16	51.74	54.01	55.49	56.63	58.39	59.76
66	39.89	42.39	43.58	46.74	48.30	50.03	52.64	54.92	56.42	57.57	59.34	60.72
67	40.66	43.18	44.39	47.58	49.16	50.90	53.53	55.83	57.34	58.50	60.29	61.68
68	41.44	43.98	45.20	48.42	50.01	51.77	54.42	56.75	58.27	59.44	61.24	62.64
69	42.21	44.78	46.02	49.26	50.87	52.65	55.32	57.66	59.20	60.37	62.19	63.60
70	42.99	45.58	46.83	50.10	51.73	53.52	56.21	58.57	60.12	61.31	63.14	64.56
71	43.77	46.39	47.64	50.95	52.59	54.39	57.11	59.49	61.05	62.25	64.09	65.52
72	44.55	47.19	48.46	51.79	53.45	55.27	58.01	60.41	61.98	63.18	65.04	66.48
73	45.33	48.00	49.28	52.64	54.31	56.14	58.90	61.32	62.91	64.12	65.99	67.44
74	46.11	48.81	50.10	53.49	55.17	57.02	59.80	62.24	63.84	65.06	66.94	68.40
75	46.90	49.61	50.92	54.34	56.03	57.90	60.70	63.16	64.76	66.00	67.89	69.37
76	47.68	50.42	51.74	55.19	56.89	58.78	61.60	64.07	65.69	66.94	68.85	70.33
77	48.47	51.23	52.56	56.04	57.76	59.65	62.50	64.99	66.63	67.88	69.80	71.29
78	49.26	52.05	53.38	56.89	58.62	60.53	63.40	65.91	67.56	68.82	70.75	72.25
79	50.05	52.86	54.21	57.74	59.49	61.41	64.30	66.83	68.49	69.76	71.70	73.22

N/D	0.01%	0.05%	0.1%	0.5%	1.0%	2%	5%	10%	15%	20%	30%	40%
80	50.84	53.68	55.03	58.60	60.36	62.30	65.21	67.75	69.42	70.70	72.66	74.18
81	51.63	54.49	55.86	59.45	61.22	63.18	66.11	68.67	70.35	71.64	73.61	75.14
82	52.43	55.31	56.69	60.30	62.09	64.06	67.01	69.59	71.28	72.58	74.57	76.11
83	53.22	56.13	57.52	61.16	62.96	64.94	67.92	70.52	72.22	73.52	75.52	77.07
84	54.02	56.95	58.35	62.02	63.83	65.83	68.82	71.44	73.15	74.46	76.47	78.04
85	54.81	57.77	59.18	62.88	64.70	66.71	69.73	72.36	74.08	75.40	77.43	79.00
86	55.61	58.59	60.01	63.73	65.57	67.60	70.63	73.28	75.02	76.35	78.38	79.97
87	56.41	59.41	60.84	64.59	66.45	68.48	71.54	74.21	75.95	77.29	79.34	80.93
88	57.21	60.23	61.67	65.45	67.32	69.37	72.45	75.13	76.89	78.23	80.30	81.90
89	58.02	61.06	62.51	66.32	68.19	70.26	73.35	76.06	77.82	79.18	81.25	82.86
90	58.82	61.88	63.34	67.18	69.07	71.15	74.26	76.98	78.76	80.12	82.21	83.83
91	59.62	62.71	64.18	68.04	69.94	72.04	75.17	77.91	79.69	81.06	83.16	84.79
92	60.43	63.54	65.02	68.90	70.82	72.92	76.08	78.83	80.63	82.01	84.12	85.76
93	61.23	64.36	65.86	69.77	71.70	73.81	76.99	79.76	81.57	82.95	85.08	86.73
94	62.04	65.19	66.70	70.63	72.57	74.71	77.90	80.69	82.50	83.90	86.03	87.69
95	62.85	66.02	67.54	71.50	73.45	75.60	78.81	81.61	83.44	84.84	86.99	88.66
96	63.66	66.85	68.38	72.36	74.33	76.49	79.72	82.54	84.38	85.79	87.95	89.62
97	64.47	67.69	69.22	73.23	75.21	77.38	80.63	83.47	85.32	86.74	88.91	90.59
98	65.28	68.52	70.06	74.10	76.09	78.27	81.54	84.39	86.26	87.68	89.87	91.56
99	66.09	69.35	70.90	74.97	76.97	79.17	82.46	85.32	87.20	88.63	90.82	92.53
100	66.91	70.19	71.75	75.84	77.85	80.06	83.37	86.25	88.13	89.58	91.78	93.49

References

Berglund, N. (2007). Chaînes de Markov. Course material, University of Orléans.

Bonald, T. and Feuillet, M. (2011). *Performances des réseaux et des systèmes informatiques*. Hermes-Lavoisier, Paris.

Foata, D. and Fuchs, A. (2004). *Processus de Poisson, chaînes de Markov et martingales*. Dunod, Malakoff.

Iversen, V.B. (2001). *Teletraffic Engineering Handbook*. Technical University of Denmark, Kongens Lyngby.

Jaton, M. and Roubaty, C. (2009). *Réseaux de télécommunications*, 7th edition. Haute École d'Ingénierie et de Gestion du Canton de Vaud, Canton de Vaud.

Leijon, H. (n.d.a). Théorie pour les groupes à accessibilité totale, Système avec attente. International Telecommunication Union, De TETRAPRO, PLATINU-Doc-31-F.

Leijon, H. (n.d.b). Théorie pour les groupes à accessibilité totale, Système avec perte. International Telecommunication Union, De TETRAPRO, PLATINU-Doc-29-F.

Petitot, M. (2010). Introduction à la modélisation des réseauux. Course material, ENIC, Villeneuve-d'Ascq.

Ravaliminoarimalalason, T.B. (2013). Etude du réseau de Whittle et son application pour modéliser les réseaux à commutation de packets. Study, DEA Sciences Cognitives et Applications, ESPA, University of Antananarivo.

Robert, P. (2000). *Réseaux et files d'attente : méthodes probabilistes*. Springer, New York.

UIT (2007). Termes et définitions relatifs à l'ingénierie du traffic. Republication of recommendation E600 from the CCITT.

Index

A, B

C, D

E

G, H, I

J, K, L, M

N, P, Q

R, S

T, W

Other titles from

iSTE

in

Networks and Telecommunications

2022

BANNOUR Fetia, SOUIHI Sami, MELLOUK Abdelhamid
Software-Defined Networking: Extending SDN Control to Large-Scale Networks (New Generation Networks SET – Volume 2)

BENAROUS Leila, BITAM Salim, MELLOUK Abdelhamid
Security in Vehicular Networks: Focus on Location and Identity Privacy (New Generation Networks Set – Volume 1)

2021

LAUNAY Frédéric
NG-RAN and 5G-NR: 5G Radio Access Network and Radio Interference

2020

PUJOLLE Guy
Software Networks: Virtualization, SDN, 5G and Security (2nd edition revised and updated)
(Advanced Network Set – Volume 1)

GONTRAND Christophe
Digital Communication Techniques

2019

LAUNEY Frédéric, PEREZ André
LTE Advanced Pro: Towards the 5G Mobile Network
Harmonic Concept and Applications

TOUNSI Wiem
Cyber-Vigilance and Digital Trust: Cyber Security in the Era of Cloud Computing and IoT

2018

ANDIA Gianfranco, DURO Yvan, TEDJINI Smail
Non-linearities in Passive RFID Systems: Third Harmonic Concept and Applications

BOUILLARD Anne, BOYER Marc, LE CORRONC Euriell
Deterministic Network Calculus: From Theory to Practical Implementation

LAUNAY Frédéric, PEREZ André
LTE Advanced Pro: Towards the 5G Mobile Network

PEREZ André
Wi-Fi Integration to the 4G Mobile Network

2017

BENSLAMA Malek, BENSLAMA Achour, ARIS Skander
Quantum Communications in New Telecommunications Systems

HILT Benoit, BERBINEAU Marion, VINEL Alexey, PIROVANO Alain
Networking Simulation for Intelligent Transportation Systems: High Mobile Wireless Nodes

LESAS Anne-Marie, MIRANDA Serge
The Art and Science of NFC Programming
(Intellectual Technologies Set – Volume 3)

2016

AL AGHA Khaldoun, PUJOLLE Guy, ALI-YAHIYA Tara
Mobile and Wireless Networks
(Advanced Network Set – Volume 2)

BATTU Daniel
Communication Networks Economy

BENSLAMA Malek, BATATIA Hadj, MESSAI Abderraouf
Transitions from Digital Communications to Quantum Communications: Concepts and Prospects

CHIASSERINI Carla Fabiana, GRIBAUDO Marco, MANINI Daniele
Analytical Modeling of Wireless Communication Systems
(Stochastic Models in Computer Science and Telecommunication Networks Set – Volume 1)

EL FALLAH SEGHROUCHNI Amal, ISHIKAWA Fuyuki, HÉRAULT Laurent, TOKUDA Hideyuki
Enablers for Smart Cities

PEREZ André
VoLTE and ViLTE

2015

BENSLAMA Malek, BATATIA Hadj, BOUCENNA Mohamed Lamine
Ad Hoc Networks Telecommunications and Game Theory

BENSLAMA Malek, KIAMOUCHE Wassila, BATATIA Hadj
Connections Management Strategies in Satellite Cellular Networks

BERTHOU Pascal, BAUDOIN Cédric, GAYRAUD Thierry, GINESTE Matthieu
Satellite and Terrestrial Hybrid Networks

CUADRA-SANCHEZ Antonio, ARACIL Javier
Traffic Anomaly Detection

LE RUYET Didier, PISCHELLA Mylène
Digital Communications 1: Source and Channel Coding

PEREZ André
LTE and LTE Advanced: 4G Network Radio Interface

PISCHELLA Mylène, LE RUYET Didier
Digital Communications 2: Digital Modulations

2014

ANJUM Bushra, PERROS Harry
Bandwidth Allocation for Video under Quality of Service Constraints

BATTU Daniel
New Telecom Networks: Enterprises and Security

BEN MAHMOUD Mohamed Slim, GUERBER Christophe, LARRIEU Nicolas, PIROVANO Alain, RADZIK José
Aeronautical Air−Ground Data Link Communications

BITAM Salim, MELLOUK Abdelhamid
Bio-inspired Routing Protocols for Vehicular Ad-Hoc Networks

CAMPISTA Miguel Elias Mitre, RUBINSTEIN Marcelo Gonçalves
Advanced Routing Protocols for Wireless Networks

CHETTO Maryline
Real-time Systems Scheduling 1: Fundamentals
Real-time Systems Scheduling 2: Focuses

EXPOSITO Ernesto, DIOP Codé
Smart SOA Platforms in Cloud Computing Architectures

MELLOUK Abdelhamid, CUADRA-SANCHEZ Antonio
Quality of Experience Engineering for Customer Added Value Services

OTEAFY Sharief M.A., HASSANEIN Hossam S.
Dynamic Wireless Sensor Networks

PEREZ André
Network Security

PERRET Etienne
Radio Frequency Identification and Sensors: From RFID to Chipless RFID

REMY Jean-Gabriel, LETAMENDIA Charlotte
LTE Standards
LTE Services

TANWIR Savera, PERROS Harry
VBR Video Traffic Models

VAN METER Rodney
Quantum Networking

XIONG Kaiqi
Resource Optimization and Security for Cloud Services

2013

ASSING Dominique, CALÉ Stéphane
Mobile Access Safety: Beyond BYOD

BEN MAHMOUD Mohamed Slim, LARRIEU Nicolas, PIROVANO Alain
Risk Propagation Assessment for Network Security: Application to Airport Communication Network Design

BERTIN Emmanuel, CRESPI Noël
Architecture and Governance for Communication Services

BEYLOT André-Luc, LABIOD Houda
Vehicular Networks: Models and Algorithms

BRITO Gabriel M., VELLOSO Pedro Braconnot, MORAES Igor M.
Information-Centric Networks: A New Paradigm for the Internet

DEUFF Dominique, COSQUER Mathilde
User-Centered Agile Method

DUARTE Otto Carlos, PUJOLLE Guy
Virtual Networks: Pluralistic Approach for the Next Generation of Internet

FOWLER Scott A., MELLOUK Abdelhamid, YAMADA Naomi
LTE-Advanced DRX Mechanism for Power Saving

JOBERT Sébastien *et al.*
Synchronous Ethernet and IEEE 1588 in Telecoms: Next Generation Synchronization Networks

MELLOUK Abdelhamid, HOCEINI Said, TRAN Hai Anh
Quality-of-Experience for Multimedia: Application to Content Delivery Network Architecture

NAIT-SIDI-MOH Ahmed, BAKHOUYA Mohamed, GABER Jaafar, WACK Maxime
Geopositioning and Mobility

PEREZ André
Voice over LTE: EPS and IMS Networks

2012

AL AGHA Khaldoun
Network Coding

BOUCHET Olivier
Wireless Optical Communications

DECREUSEFOND Laurent, MOYAL Pascal
Stochastic Modeling and Analysis of Telecoms Networks

DUFOUR Jean-Yves
Intelligent Video Surveillance Systems

EXPOSITO Ernesto
Advanced Transport Protocols: Designing the Next Generation

JUMIRA Oswald, ZEADALLY Sherali
Energy Efficiency in Wireless Networks

KRIEF Francine
Green Networking

PEREZ André
Mobile Networks Architecture

2011

BONALD Thomas, FEUILLET Mathieu
Network Performance Analysis

CARBOU Romain, DIAZ Michel, EXPOSITO Ernesto, ROMAN Rodrigo
Digital Home Networking

CHABANNE Hervé, URIEN Pascal, SUSINI Jean-Ferdinand
RFID and the Internet of Things

GARDUNO David, DIAZ Michel
Communicating Systems with UML 2: Modeling and Analysis of Network Protocols

LAHEURTE Jean-Marc
Compact Antennas for Wireless Communications and Terminals: Theory and Design

PALICOT Jacques
Radio Engineering: From Software Radio to Cognitive Radio

PEREZ André
IP, Ethernet and MPLS Networks: Resource and Fault Management

RÉMY Jean-Gabriel, LETAMENDIA Charlotte
Home Area Networks and IPTV

TOUTAIN Laurent, MINABURO Ana
Local Networks and the Internet: From Protocols to Interconnection

2010

CHAOUCHI Hakima
The Internet of Things

FRIKHA Mounir
Ad Hoc Networks: Routing, QoS and Optimization

KRIEF Francine
Communicating Embedded Systems / Network Applications

2009

CHAOUCHI Hakima, MAKNAVICIUS Maryline
Wireless and Mobile Network Security

VIVIER Emmanuelle
Radio Resources Management in WiMAX

2008

CHADUC Jean-Marc, POGOREL Gérard
The Radio Spectrum

GAÏTI Dominique
Autonomic Networks

LABIOD Houda
Wireless Ad Hoc and Sensor Networks

LECOY Pierre
Fiber-optic Communications

MELLOUK Abdelhamid
End-to-End Quality of Service Engineering in Next Generation Heterogeneous Networks

PAGANI Pascal *et al.*
Ultra-wideband Radio Propagation Channel

2007

BENSLIMANE Abderrahim
Multimedia Multicast on the Internet

PUJOLLE Guy
Management, Control and Evolution of IP Networks

SANCHEZ Javier, THIOUNE Mamadou
UMTS

VIVIER Guillaume
Reconfigurable Mobile Radio Systems

Printed and bound by CPI Group (UK) Ltd, Croydon, CR0 4YY
13/09/2026
14961327-0001